T0074132

THE FRONTIERS COLLECTION

THE FRONTIERS COLLECTION

Series Editors:
A.C. Elitzur L. Mersini-Houghton M.A. Schlosshauer M.P. Silverman
J.A. Tuszynski R. Vaas H.D. Zeh

The books in this collection are devoted to challenging and open problems at the forefront of modern science, including related philosophical debates. In contrast to typical research monographs, however, they strive to present their topics in a manner accessible also to scientifically literate non-specialists wishing to gain insight into the deeper implications and fascinating questions involved. Taken as a whole, the series reflects the need for a fundamental and interdisciplinary approach to modern science. Furthermore, it is intended to encourage active scientists in all areas to ponder over important and perhaps controversial issues beyond their own speciality. Extending from quantum physics and relativity to entropy, consciousness and complex systems – the Frontiers Collection will inspire readers to push back the frontiers of their own knowledge.

Other Recent Titles

The Biological Evolution of Religious Mind and Behavior
By E. Voland, W. Schiefenhövel

Entanglement, Information, and the Interpretation of Quantum Mechanics
By G. Jaeger

Homo Novus - A Human Without Illusions
By U.J. Frey, C. Störmer, K.P. Willführ

The Physical Basis of the Direction of Time
By H.D. Zeh

Mindful Universe
Quantum Mechanics and the Participating Observer
By H. Stapp

Decoherence and the Quantum-To-Classical Transition
By M.A. Schlosshauer

The Nonlinear Universe
Chaos, Emergence, Life
By A. Scott

Symmetry Rules
How Science and Nature Are Founded on Symmetry
By J. Rosen

Quantum Superposition
Counterintuitive Consequences of Coherence, Entanglement, and Interference
By M.P. Silverman

For all volumes see back matter of the book

**Ulrich J. Frey · Charlotte Störmer ·
Kai P. Willführ**

Editors

ESSENTIAL
BUILDING BLOCKS OF
HUMAN NATURE

 Springer

Editors

Ulrich J. Frey
University of Giessen
Center for Philosophy and
Foundations of Science
Otto-Behaghel-Str. 10
35394 Giessen
Germany
Ulrich.frey@phil.uni-giessen.de

Charlotte Störmer
University of Giessen
Center for Philosophy and
Foundations of Science
Otto-Behaghel-Str. 10
35394 Giessen
Germany
Charlotte.Stoermer@phil.
uni-giessen.de

Kai P. Willführ
University of Giessen
Center for Philosophy and
Foundations of Science
Otto-Behaghel-Str. 10
35394 Giessen
Germany
Kai.P.Willfuehr@phil.
uni-giessen.de

Series Editors:

Avshalom C. Elitzur
Bar-Ilan University, Unit of Interdisciplinary Studies, 52900 Ramat-Gan, Israel
email: avshalom.elitzur@weizmann.ac.il

Laura Mersini-Houghton
Dept. Physics, University of North Carolina, Chapel Hill, NC 27599-3255, USA
email: mersini@physics.unc.edu

Maximilian A. Schlosshauer
Niels Bohr Institute, Blegdamsvej 17, 2100 Copenhagen, Denmark
email: schlosshauer@nbi.dk

Mark P. Silverman
Trinity College, Dept. Physics, Hartford CT 06106, USA
email: mark.silverman@trincoll.edu

Jack A. Tuszynski
University of Alberta, Dept. Physics, Edmonton AB T6G 1Z2, Canada
email: jtus@phys.ualberta.ca

Rüdiger Vaas
University of Giessen, Center for Philosophy and Foundations of Science, 35394 Giessen, Germany
email: ruediger.vaas@t-online.de

H. Dieter Zeh
Gaiberger Straße 38, 69151 Waldhilsbach, Germany
email: zeh@uni-heidelberg.de

Corrected 2nd Printing 2011

ISSN 1612-3018
ISBN 978-3-642-13967-3   e-ISBN 978-3-642-13968-0
DOI 10.1007/978-3-642-13968-0
Springer Heidelberg Dordrecht London New York

Library of Congress Control Number: 2010938112

Cover design: KuenkelLopka GmbH, Heidelberg

Printed on acid-free paper

Springer is part of Springer Science+Business Media (www.springer.com)

To Eckart Voland

Preface

Why do we like what we like? Why do we believe in the supernatural? Where do our mathematical abilities come from? These questions form parts of the overarching question of what constitutes human nature. Any attempt to understand human nature has to take knowledge from different fields. Ultimately, as Kant famously pointed out, they converge in: What is Man?

Humans have always reflected not only on nature, but on their very own nature — such questions therefore date back to the roots of scientific research, and in their classical form can already be found in Greek philosophy, 2300 years ago. Since then, different ages have found different answers. However, there has been one theory — Darwinian theory — which has succeeded in providing satisfactory explananations for diverse, formerly disparate fields, thereby uniting them within one framework.

So even though the questions remain the same, the new and sometimes surprising answers found by evolutionary theory are very different from those of Plato or Aristotle. In this sense, Darwinian theory provides a grounding for a broad range of researchers and sheds new light on fields of enquiry as old as aesthetics and religion.

Interestingly, there are few attempts to go beyond the use of Darwinian theory as a methodological tool. Fewer still try to unite isolated breakthroughs achieved by this paradigm into one coherent picture and theory. The aim of this book is to encourage just such an enterprise.

We would like to thank our mentor Eckart Voland for inspiring and teaching us to think in a genuinely interdisciplinary way. Furthermore, we would like say thank you to all authors who added their own piece to the mosaic we had in mind with this volume. Last, but not least, we appreciate the helpfulness and encouragement of Angela Lahee.

Giessen, Germany,   *Ulrich J. Frey, Charlotte Störmer, Kai P. Willführ*
August 2010

Contents

List of Contributors

Ulrich J. Frey
University of Giessen, Zentrum für Philosophie und Grundlagen der Wissenschaft,
Otto-Behaghel-Str. 10C, 35394 Giessen, Germany, e-mail: ulrich.frey@phil.uni-giessen.de

Karl Grammer
Department of Anthropology, University of Vienna, Althanstrasse 14, 1090 Vienna,
Austria, e-mail: urban.ethology@univie.ac.at

Peter M. Kappeler
Abteilung für Verhaltensökologie & Soziobiologie, Deutsches Primatenzentrum,
Kellnerweg 4, 37077 Göttingen, Germany, e-mail: pkappel@gwdg.de

Niklas Krebs
University of Giessen, Zentrum für Philosophie und Grundlagen der Wissenschaft,
Otto-Behaghel-Str. 10C, 35394 Giessen, Germany, e-mail: Krebs-Homebase@t-online.de

Ruth Mace
Department of Anthropology, University College London, Gower Street, London
WC1E 6BT, United Kingdom, e-mail: r.mace@ucl.ac.uk

Elisabeth Oberzaucher
Department of Anthropology, University of Vienna, Althanstrasse 14, 1090 Vienna,
Austria, e-mail: elisabeth.oberzaucher@univie.ac.at

Benjamin G. Purzycki
Department of Anthropology, University of Connecticut, 354 Mansfield Road,
Storrs, CT, USA, e-mail: benjamin.purzycki@uconn.edu

Frank Rösler
Institute of Psychology, University of Potsdam, Karl-Liebknecht-Str. 24/25, 14476
Potsdam OT Golm, Germany, e-mail: froesler@uni-potsdam.de

Michael Schmidt-Salomon
c/o GBS-Bro Elke Held, Im Gemeindeberg 21, 54309 Besslich, Germany, e-mail:
mss@schmidt-salomon.de, homepage: www.schmidt-salomon.de/

Mary K. Shenk
Department of Anthropology, University of Missouri, 107 Swallow Hall, Columbia,
MO 65211-1440, USA, e-mail: shenkm@missouri.edu

Richard Sosis
Department of Anthropology, University of Connecticut, 354 Mansfield Road,
Storrs, CT, USA, e-mail: richard.sosis@uconn.edu

Charlotte Störmer
University of Giessen, Zentrum für Philosophie und Grundlagen der
Wissenschaft, Otto-Behaghel-Str. 10C, 35394 Giessen, Germany, e-mail:
charlotte.stoermer@phil.uni-giessen.de

Matthias Uhl
Phönixstr. 3, 35578 Wetzlar, Germany, e-mail: m@tthiasuhl.de, homepage:
www.matthias-uhl.de

Kai P. Willführ
University of Giessen, Zentrum für Philosophie und Grundlagen der
Wissenschaft, Otto-Behaghel-Str. 10C, 35394 Giessen, Germany, e-mail:
kai.p.willfuehr@phil.uni-giessen.de

Introduction

Ulrich J. Frey, Charlotte Störmer, and Kai P. Willführ

Probably the best way to understand a complex phenomenon is to decompose it into its components. In particular, this holds true for the undoubtedly complex issue of human nature. Each component requires its own discipline, and the challenge is to recombine the results into a meaningful whole. The picture we have in mind is that of a mosaic, where single pieces produce a picture. Of course, the motif of the mosaic — human nature — cannot be seen until most of the pieces have been fitted together. Unfortunately, scientific work more closely resembles a mosaic than a jigsaw, as the edges/borders are largely unknown. Besides, the interdisciplinary work required is not standard practice. We call the tiles of the mosaic 'building blocks', attempting to glimpse the motif of human nature in this volume by assembling the ideas of leading authors from different disciplines. Each individual result contributes in a unique way to the overall picture.

Complicating research into human nature is the fact that its building blocks are not mutually independent. The structure of the book is a direct consequence of this. It is an attempt to identify the essential building blocks of human nature and to fit them together in a meaningful way. Essential for the problem of human nature are not only primatology and anthropology, describing fundamental distinctions between humans and our nearest relatives, the great apes, but also sociobiology and its associated fields. They explain subtle differences in human behavior, regarding

Ulrich J. Frey
University of Giessen, Zentrum für Philosophie und Grundlagen der Wissenschaft, Otto-Behaghel-Str. 10C, 35394 Giessen, Germany,
e-mail: ulrich.frey@phil.uni-giessen.de

Charlotte Störmer
University of Giessen, Zentrum für Philosophie und Grundlagen der Wissenschaft, Otto-Behaghel-Str. 10C, 35394 Giessen, Germany,
e-mail: charlotte.stoermer@phil.uni-giessen.de

Kai P. Willführ
University of Giessen, Zentrum für Philosophie und Grundlagen der Wissenschaft, Otto-Behaghel-Str. 10C, 35394 Giessen, Germany,
e-mail: kai.p.willfuehr@phil.uni-giessen.de

for example the way humans cooperate, or the way emotions trigger decisions. Richly diversified fields flow from these, encompassing aesthetic perception and cultural and media studies. The ability of humans to engage in finely grained activities like philosophy and mathematics can be explained by the aforementioned building blocks.

The following contributions help us to assemble this jigsaw. In the first chapter of the book, Peter Kappeler discusses the phylogenetic roots of human behavior. He demonstrates that there are shared behavioral similarities between humans and other primates, a thesis founded on the high degree of genetic and anatomic similarity. But beyond these shared traits, humans are unique in their prosocial disposition. This enables us to reach high levels of social learning, communication, and cooperation.

The second chapter, by Mary K. Shenk, is an introduction to parental investment theory in humans. Starting from the concept of reproductive tradeoffs (when to reproduce and how many offspring to have), she outlines the importance of helping kin, differential parental investment in offspring, and sibling competition for parental resources, all of these being determinant factors in reproductive success.

Ruth Mace then tests adaptive hypotheses about behavioral and cultural diversity in Chap. 3. Using cross-cultural comparisons, she emphasizes ecological correlates of behavior. In particular, she argues that the subsistence system is a key factor shaping both human biology and social structure.

In Chap. 4, Frank Rösler examines results regarding the neuronal bases of decision-making. These findings suggest that there exists a hierarchical architecture with central selection switches. This in turn suggests that selections and decisions are fully deterministic from a biological perspective.

There has been a surge of interest in evolutionary studies of religion over the past few years. Benjamin Purzycki and Richard Sosis discuss the question of supernatural minds and their cross-cultural variation in Chap. 5. They argue that the social or cultural orientation of religions depends on different needs, and analyse religious agency and its attributes.

Karl Grammer and Elisabeth Oberzaucher discuss the fascinating question of evolutionary aesthetics in Chap. 6. What is beautiful for us and what is not? They describe the 'eight pillars of attractiveness' and show how we implement a mechanism for avoiding ugliness. On the basis of this evidence, the authors then go on to ask how such mechanisms will fare in the future, particularly under the influence of the modern media.

But there is a sense in which today's media are not as innovative as they may appear to be. This is the message of Matthias Uhl's contribution in Chap. 7. He contends that the modern media appeal to basic and indeed ancient mechanisms of our evolved brain. We are still reacting to the same cues as did our ancestors. Thus, we are often unaware of the ways the modern media are able to manipulate us. For example, an analysis of the main plot elements in both Hollywood and Bollywood movies shows that things like mate selection or dangerous situations are the most prominent themes in the majority of such films, regardless of the cultural background and their actual frequency in real life.

Niklas Krebs' topic in Chap. 8 is the evolutionary foundations of our mathematical abilities. He argues that the roots of counting are to be found in the assessment of quantity, and the roots of mathematical relations in our complex social systems. His work presented here may well be one of the first evolutionary accounts of mathematical abilities.

In the final chapter, after a sharp critique of creationism and its modern variant intelligent design, Schmidt-Salomon outlines evolutionary humanism as a new world view compatible with evolutionary theory. He clarifies misunderstandings commonly associated with this theory, and goes on to make a strong case for it.

To sum up then, we hope to have brought together enough expertise throughout this volume to provide a glimpse of an overall pattern that may be better assessed when we finally assemble the jigsaw of human nature.

Chapter 1
Our Origins: How and Why We Do and Do Not Differ from Primates

Peter Kappeler

Abstract Questions about human origins and uniqueness are at the core of unraveling the essential building blocks of human nature. Probably no other single topic has received more attention across the sciences and humanities than the question of what makes us human and how humans differ from other primates and animals. Evolutionary anthropologists can contribute important comparative evidence to this debate because they adopt a broad perspective that considers both the ancestors of the human species as well as its closest living biological relatives. In this chapter, I review some recent insights into human nature based on this perspective. My focus is on social behavior and its underlying adaptations and mechanisms, because this is the realm of man's most salient features. In contrast to many mainstream contributions on this topic, I emphasize shared behavioral similarities between humans and other primates and outline their underlying mechanisms. These behavioral features shared with other primates include much of our homeostatic behavior and many of our emotions and cognitive abilities, so that together they appear to represent the submerged part of an iceberg. I also briefly summarize some of the uniquely human traits forming the tip of the iceberg and outline current attempts to explain their origin. Accordingly, in this context shared intentionality represents a crucial psychological mechanism that may have been reinforced by a switch to a cooperative breeding system in early *Homo* evolution. In conclusion, this essay contends that the key essential building block defining human nature is like the core of a Russian doll, while all the outer layers represent our vertebrate, mammalian, and primate legacies.

Peter Kappeler
Abteilung für Verhaltensökologie & Soziobiologie, Deutsches Primatenzentrum, Kellnerweg 4, 37077 Göttingen, Germany, e-mail: pkappel@gwdg.de

U.J. Frey et al. (eds.), *Essential Building Blocks of Human Nature*, The Frontiers Collection, DOI 10.1007/978-3-642-13968-0_1, © Springer-Verlag Berlin Heidelberg 2011

1.1 Introduction

The question as to how and why humans differ from non-human primates and other animals has preoccupied philosophers and theologians, anthropologists and biologists, and human and social scientists for millennia (Hill et al. 2009; Kappeler et al. 2010). Whereas historical attempts to answer this question tended to focus on single, absolute criteria, such as language, warfare, or tool use, contemporary scholars are beginning to appreciate the contributions from evolutionary anthropologists over the past 50 years or so. These have yielded important insights into both the lives and habits of our hominid ancestors, as well as into the social behavior and cognitive abilities of living non-human primates. This body of research has begun to sketch the contours of a continuum in a number of relevant traits that transcend both time and species boundaries. Thus, by trying to pinpoint the degree to which *Homo sapiens sapiens* differs from both its immediate ancestors and its closest living biological relatives (including primates other than members of the genus *Pan*), a much more differentiated answer can be found. Moreover, such a broad comparative approach not only highlights differences between taxa, but can identify some traits as the result of shared common descent (e.g., Fichtel & Kappeler 2010; Whiten 2010). In this essay, I will review and highlight some recent studies of human universals (in the sense of traits unique to our species) together with some attempts to identify shared basal features that firmly link humans to other primates. I will also summarize some reasons for these similarities and differences. Because of the nature and scope of the topic, this review will have to be superficial and eclectic, and in contrast to most other contributions on this topic, my emphasis will be on the behavioral symplesiomorphies of humans.

Because Carolus Linnaeus shied away from a formal diagnosis in his scientific description of humans, it was left to later scientists to identify the anatomical synapomorphies that distinguish *Homo sapiens* from the other hominids. The list is surprisingly short and includes mostly anatomical traits such as bipedalism (and functionally related adaptations), lack of an opposable big toe, an enlarged neocortex, and permanent breasts (Lovejoy 1981). Similarly, the analysis of the hominid fossil record provides a rough outline of the timing and sequence of anatomical changes leading up to the emergence of *Homo sapiens* about 160 000 years ago, but most of the details have to do with bipedalism and changes in dentition and cranial volume, or they reflect changes in degree (e.g., in skull shape or brain size) rather than fundamental innovations (Henke & Tattersall 2007). Thus, comparative anatomists and paleoanthropologists can clearly identify a human being and distinguish it unequivocally from our closest biological relatives in the present or past, but their list of criteria does not answer the fundamental question about human nature in a manner that would deeply satisfy scholars of most other disciplines.

Since humans do not differ qualitatively in their anatomy from great apes, except for the adaptations related to bipedalism, because their life histories are broadly similar to each other (menopause notwithstanding) and because the genetic differences between humans and great apes lie in the lower range of established species differences, the main difference must exist in the realm of behavior and cognition. There is indeed little doubt that *Homo sapiens* is the most intelligent and socially complex animal. Human cultural and technological achievements, powered by our large brains and capacity for language, are impressive. Within a few thousand generations, we have changed from hunter–gatherers and agro-pastoralists to post-industrialists that build spacecraft, fiddle with nuclear power, manipulate the genome of other creatures, cure and eradicate diseases, and transmit information instantaneously around the globe with telephones and computers. It is widely accepted that our intelligence and rationality are the salient driving forces of human behavior which facilitate all those achievements. However, the very same rational individuals engage in futile contests over social status, discriminate against members of other social groups, and exhibit jealousy and vanity. And at the same time humans also donate money to support common welfare, help strangers, promote their kin, respond in predictable ways to particular stimuli of beauty or emotion, and consistently exhibit sex differences in many aspects of social behavior across cultures. Evolutionary processes, therefore, have also profoundly shaped some patterns of human social organization and behavior. The exact mechanisms by which evolutionary and cultural mechanisms interact in shaping human social behavior is still being discovered (McElreath 2010). What is clear, however, is that the salient human universals are behavioral and cognitive; what is less clear is which traits are really unique and what makes them unique. The answer to this question may benefit from an explicit consideration of symplesiomorphies, i.e., traits found in two or more taxa because they were inherited from a common ancestor.

1.2 Why Humans Do Not Differ from Primates

The vast majority of us have a clear vision or memory of both our parents. Most of us are also fortunate enough to have shared part of our lives with our grandparents. However, personal acquaintance with great-grandparents is already a very rare event. If you are willing to embark on virtual time travel, you can extend this mental game by imagining what your grandparents' grandparents looked like and what their living conditions may have been like. From there, you can go back another 10, 100, 1000, or even more generations. Because every single part of your family tree is dependent on the successful survival and reproduction of every single one of the dozens, hundreds, thousands, or millions of direct ancestors, it is intuitively easy to capture and appreciate the concept of biological continuity, and this explains why we share particular traits and with which other animals.

At least since Carolus Linnaeus placed *Homo sapiens* within the order of primates, biologists and anthropologists have acknowledged that humans share a number of physical similarities with other animals. Accordingly, the combination of morphological, physiological, and behavioral traits that characterize modern humans as a primate species is the result of biological continuity over the millennia on the one hand, and the more recent acquisition of species-specific traits on the other. The tiered hierarchy of these similarities reflects the fact that we are members of the natural biological hierarchy, where taxa with a relatively recent common ancestor are more likely to share more traits with each other than species with more distant common ancestors. Thus, *Homo sapiens* has a backbone, for example, because we are members of the class of vertebrates. We have hair because we are mammals, we have nails instead of claws on our digits because we are primates, we have a dry, mobile upper lip because we are haplorrhine primates, our nostrils face forward and downward because we are catarrhine primates, we lack a tail because we are hominoids, we have relatively broad incisors because we are hominids, and we are bipedal because we share this trait with all species of the genus *Homo*, while we have a steeper forehead and a much more pronounced chin than other members of our genus. Thus, in a way, human biology resembles a Russian doll, where only features of the smallest doll represent newly acquired traits and all the outer ones represent legacies of past phylogenetic stages. A similar list of traits could be assembled for physiological traits, although it would require a journey much deeper into time to identify the origins of many physiological processes that provide our bodies with energy or furnish our sensory organs with their particular functions.

The presence of such primitive characters shared with other species and taxa is ultimately due to the effects of genes and their continuous transmission. The information encoded in the sequence of DNA bases provides the blueprint for assembling proteins, which, in turn, make up our cells and morphological characters or control and coordinate their assembly from other basic materials. Every zygote requires this genetic information from its parental germ cells in order to develop properly and to grow into a functioning organism. At every speciation event, a new lineage acquires, by definition, some new heritable traits that distinguish it from its ancestor, while maintaining the majority of traits that have stood the test of time and enabled all of its ancestors to survive and reproduce successfully. As a result, the similarity in DNA sequences between a given pair of species is a function of the time since they shared their last common ancestor. Thus, humans and mice share about 80% of their DNA sequences because only about 20% of the genetic information was modified in both lineages leading to modern mice and humans in the 90 million years since these species shared their last common ancestor (Church et al. 2009). Similarly, in the 5 million years since the lineages leading to modern humans and chimpanzees separated, only about 1.2% of their genetic material has been modified (Mikkelsen et al. 2005), and a disproportionately large share of these genes is functionally related to neural development and function (as well as nutritional modifications) (Haygood et al. 2007). Thus, our bodies and genes provide unequivocal evidence for the fact that humans are vertebrates, mammals, primates, hominids, etc. This insight should be unsurprising for all but hardcore creationists. A much more interesting question arises with respect to the genetic basis of another set of traits that also characterize and distinguish species: their behavior.

Comparison of the human and chimpanzee genome suggests that everything that distinguishes us from chimpanzees must be encoded in the very small amount of uniquely human DNA. This hypothesis is based on the assumption that all morphological, physiological, and behavioral traits are controlled by the genes that we can sequence. With respect to behavior, this explanation could only be correct if relatively small genetic differences corresponded to major behavioral differences. So far, however, students of animal behavior have identified only intraspecific examples of gene-dependent behavioral phenotypes (Taborsky & Brockman 2010). In voles (*Microtus* spp.), for example, relatively minor genetic differences in a vasopressin receptor gene correspond to major species differences in social organization and the mating system (Hammock & Young 2005). However, a recent study of the same vasopressin receptor gene in 12 Old World primates with variable mating systems revealed no co-variation (Rosso et al. 2008), indicating that there is probably no invariant 1:1 relationship between the particular version of a gene and a predictable behavioral outcome. Behavior is simply shaped and constrained by too many other factors (Kappeler & Kraus 2010).

Another potential example for minor genetic differences with major behavioral consequences, in this case for humans, is provided by the FOXP2 gene, which is involved in the control of the neural circuitry handling speech and language (Vargha-Khadem et al. 2005). Its present form in humans, which differs from that of other great apes by only a few mutations, has been present for about 200 000 years, roughly coinciding with the emergence of modern humans (Enard et al. 2002). On the other hand, mice and humans also share more than 93% of their FOXP2 sequences (ibid.), but mice do not talk. Thus, a detailed molecular understanding of single amino acid substitutions is required to functionally link genetic differences to behavioral ones (Marcus & Fisher 2003), so that the function and action of a single candidate gene can be illuminated. Similarly, *Microcephalin*, a gene involved in the regulation of brain growth, is more variable in humans than in other primates. It has been under positive selection since the origin of the last common ancestor of humans and great apes (Wang & Su 2004), and one genetic variant of this gene in humans has been under positive selection in the past 40 000 years (Evans et al. 2005). Thus, some important autapomorphies of humans related to our behavior and cognition, such as language and enlarged brains, appear to have a genetic underpinning, but, crucially, it remains largely unclear which aspects of human behavior are under direct genetic control and to what extent, and in particular, we do not know the mediating mechanisms.

The main fundamental problem in this context is that we lack a general understanding of how gene sequences are translated into behavior. Even in invertebrates with relatively simple central nervous systems and unlimited opportunities for invasive research, only the two ends of a black box, i.e., changes in gene activity and in behavior, can (currently) be correlated with each other (e.g. Whitfield et al. 2003; Cirelli et al. 2005; Orr 2005; Dickson 2008; Price et al. 2008; Takahasi et al. 2008), and we do not know how particular gene products proximately elicit a particular behavior via the underlying neural networks; in fact, even the simulation of relatively simple neural networks continues to pose massive computational challenges (Markram 2006), so we are currently unable to explain how neuronal activity in vertebrates generates simple behavior, let alone complex social behavior patterns (Robinson et al. 2008). In addition, a developing fetus and juvenile is exposed to a constant array of stimuli that can have lasting effects on their behavior as adults (Sachser & Kaiser 2010). It is therefore highly questionable whether a mechanistic understanding of the non-linear gene–neuron–behavior relationships in primates and humans will ever be possible.

Thus, in many respects, humans do not differ from primates and other animals because we share the same genetic and developmental blueprints that allowed our ancestors to survive and reproduce successfully. This proximate explanation applies to behavior patterns as well, even though we know next to nothing about the underlying molecular mechanisms. Some of the behavioral traits of humans that may be explained as a result of symplesiomorphies are illustrated in the following section.

1.3 How Humans Do Not Differ from Primates

Behavioral similarities between primates and humans are widespread but little recognized and appreciated. This may have to do with the fact that 'behavior' is rarely explicitly defined and most people have cognitively demanding or socially complex behavior patterns in mind when undertaking interspecific comparisons. As a result, the emphasis is therefore often on differences, rather than similarities. However, a good example for a functional context in which numerous behavior patterns are highly invariant across lineages is provided by homeostasis, which refers to a closely regulated state of several basic organismic functions. Behavioral acts and pattern play an important proximate role in the control of most of these functions. For example, like all other Old World primates, humans are characterized by diurnal activity that is deeply embedded in a circadian activity rhythm. Sleep is the main behavioral mechanism used to mediate the body's need for rest and inactivity, and it constitutes about a third of our daily behavior. The type of activity and also the control of the circadian rhythm have a strong genetic component to them (e.g., Reppert & Weaver 2002; Takahashi et al. 2008). Thirst and hunger provide other examples where behavioral actions are tightly integrated into physiological feedback loops. Thus, just like other primates and animals, we eat and drink whenever our bodies signal a need for additional energy or fluid. Thermoregulation provides a final example of behavior patterns controlled unconsciously by our brain stem that serve to maintain comfortable body temperatures. Collectively, these homeostatic behavior patterns contribute significantly to daily well-being, and hence successful survival, and these routines are so basal that they have been largely delegated to the unconscious part of our central nervous system. Interestingly, however, homeostatic behavior is rarely mentioned in sociobiological controversies by proponents of both sides of the old nature–nurture debate (e.g., Smith et al. 2001), even though the deep, hard-wired control of these behavior patterns is undisputable.

To the extent that emotions are accompanied or expressed by specific behavior patterns, they can also be said to constitute basal traits that humans share with primates and some other vertebrates. In fact, emotions are essentially adaptations for behavioral regulation that evolved in response to recurrent social or ecological challenges (Fessler & Gervais 2010). Closer inspection and comparative research has revealed a hierarchy of emotions. First, some emotions, such as fear and disgust and their behavioral manifestations appear to be subject to phylogenetically old brain regions or endocrine systems, and are executed with minimal cognitive requirements. Again, these responses appear highly adaptive and are tightly integrated with physiological systems. Their positive effects on survival or reproduction are in most cases very obvious, and it is therefore not surprising that the proximate mechanisms underlying the behavioral components are highly conserved and invariant. However, some of these emotional responses, such as fear of spiders, vertigo, or claustrophobia, can be suppressed by learning, and we do not know whether such phenomena exist among animals as well. Second, emotions such as anger, jealousy, and parental love are involved in elementary social interactions, such as competition, parenting, or pair-bonding, and hence are also likely to be shared among most primates and other mammals with the corresponding social system. Third, it can be argued that dyadic cooperative relationships are particularly important and widespread among primates (Silk 2009; Silk & Boyd 2010), and they appear to be connected to a particular set of emotions, such as gratitude and guilt, which require some theory of mind. Finally, there are vicarious emotions, elicited when events that befall a conspecific are treated as if they had befallen the self, and these require advanced cognitive abilities, so may be limited to the great apes, or even the genus *Pan* and humans (Fessler & Gervais 2010).

Some components of our cognitive abilities also have a phylogenetic legacy that is millions of years old. By studying cognitive abilities in living primates (and other animals), we can use parsimony principles to infer the presence or absence of certain traits from their distribution across a phylogenetic tree that includes their common ancestors. Accordingly, aspects of technical intelligence, including cognitive preoccupation with space, objects, features, categories, quantities, tools, and causality, have deep roots within the primate order reaching all the way down to the most basal lemurs (Fichtel & Kappeler 2010). Similarly, hallmarks of social intelligence, such as gaze-following, coalition-formation, deception, and social learning, are also present, at least in some rudimentary forms, in most primates studied to date. Detailed comparisons among primates can be found in Tomasello & Call (1997) and Fichtel & Kappeler (2010); a systematic comparison with other mammals and birds (e.g., Bugnyar & Kotrschal 2002; Clayton & Dickinson 1998; Blaisdell et al. 2006) would also be interesting.

Thus, a Russian doll principle can be discerned with respect to the evolutionary hierarchy of homeostatic behavior, emotions, and cognitive abilities, where the vast majority of items represent traits shared with other primates, mammals, or vertebrates and only some core features are uniquely human.

1.4 How Humans Do Differ from Primates

Because of my intended focus in this essay on the usually underexposed other side of the coin (i.e., how humans do not differ from other primates), I will mainly use the subsequent sections on human uniqueness to refer the interested reader to recent selected examples and summaries. First, with respect to the emotional repertoire outlined above, humans also appear to exhibit some unique features. In particular, moral emotions related to norm violations appear to be unique to humans (Fessler & Gervais 2010). Only our species has developed socially transmitted standards that define expectations and actions that promote prosociality (McElreath et al. 2003). Moral outrage and moral disgust are emotions faced by non-cooperative norm violators. Direct and altruistic punishment are important behavioral mechanisms in this context to enforce social norms (Fehr & Gächter 2002; Gintis et al. 2003). Second, humans exhibit unique cognitive abilities, including language, long-term planning, causal understanding, and episodic memory. These abilities build on shared intentionality, i.e., the ability to participate with others in collaborative activities with shared goals and intentions (see below), which also involves language-based teaching. Third, humans alone have developed cumulative material culture, social institutions, and rituals, including religion, all critically dependent on language and social learning (McElreath 2010). Across cultures, human subsistence ecology is characterized by skill-intensive hunting and gathering, sexual division of labor, and extremely intense cooperation with kin and non-kin, also in between-group conflict (Kaplan et al. 2000, Gurven 2004; Silk & Boyd 2010, Gat 2010). Finally, with respect to reproductive behavior and life history, humans differ from other primates because of their larger brains, slower development, longer lifespans, higher female reproductive rates, midlife menopause (Robson et al. 2006), and extensive allomaternal help (Hrdy 2009). The latter point plays a central role in the most recent hypothesis put forward to provide an adaptive explanation of human behavioral uniqueness.

1.5 Why Humans Do Differ from Primates

As in other areas of behavioral inquiry, 'why' questions can be answered on several levels. Apart from the phylogenetic legacy outlined above and ontogenetic approaches focusing on social learning, including teaching (Bjorklund et al. 2010; van Schaik 2010), a particular proximate behavioral mechanism and a specific adaptive scenario have featured prominently in recent discussions of human behavioral uniqueness.

First, shared intentionality provides a crucial and uniquely human psychological mechanism that has apparently facilitated the evolution of language, cumulative culture, and complex cooperation. It is based on a prosocial disposition not seen among other primates with advanced levels of technical and social intelligence. This special disposition is manifested in unique motivations and socio-cognitive skills for understanding other individuals as potential cooperators with whom mental states, such as attention, emotions, experience, and also collaborative actions can be shared (Tomasello et al. 2005). Shared intentionality therefore provides a social mechanism to connect several already relatively high-powered brains to achieve new levels of social performance. Comparative research on great apes and humans has demonstrated that various operationalizable aspects of this disposition exist solely in humans, albeit only from about 1–2 years of age on (Tomasello & Moll 2010). In contrast, chimpanzees fail to exhibit a joint commitment to a shared goal and to adopt reciprocal roles in a common task (Call & Tomasello 2008). This prosocial disposition in humans has been ratcheted up in a self-reinforcing way by social learning about how to communicate and cooperate with conspecifics, and what to learn from them. During ontogeny, mothers play a crucial role in this process (Bjorklund et al. 2010).

Second, the psychological disposition towards unusual prosociality has had important adaptive consequences in the context of human reproduction. According to a new, comprehensive hypothesis, allomothering or ('cooperative breeding') has played a central role in the evolution of human uniqueness during the early stages of the evolution of the genus *Homo* (Hrdy 2009; Burkart et al. 2009). Accordingly, in a first step, social support of pregnant and nursing mothers by fathers, grandmothers, older siblings, and other close relatives has lifted energetic limitations on brain growth and shifted human life histories toward the current pattern (van Schaik & Burkart 2010). This crucial change in the human social system may have been substantially facilitated by a change in the human mating system from chimpanzee-like promiscuity to a pronounced monogamous pattern (Chapais 2008). Cooperative breeding, in turn, requires extreme social tolerance and group-level cooperation, and, crucially, affects social cognition directly and indirectly. Most directly, it broadens opportunities for social learning and enhances social coordination, which is initially required to organize collective infant care. Indirectly, enhanced opportunities for social learning provide additional and improved conditions for implementing the cognitive abilities derived from increased brain size, leading eventually to a self-reinforcement of this social system and all its morphological, psychological, and behavioral components. Some of the assumptions and predictions of this hypothesis have enjoyed empirical support, promising a powerful explanation of the particular traits that characterize human uniqueness.

Acknowledgements I would like to thank Eckart Voland for constant inspiration over the years, and Charlotte Störmer, Kai Willführ, and Ulrich Frey for the invitation to contribute to this volume. As with so many other insights, the Russian doll principle was introduced to me by Carel van Schaik.

References

Bjorklund DF, Causey K, Periss V (2010) The evolution and development of human social cognition. In: Kappeler PM, Silk JB (eds) *Mind the Gap: Tracing the Origins of Human Universals.* Springer, Heidelberg, pp 351–371

Blaisdell AP, Sawa K, Leising KJ, Waldmann MR (2006) Causal reasoning in rats. Science 311:1020–1022

Bugnyar T, Kotrschal K (2002) Observational lerning and the raiding of food caches in ravens, *Corvus corax*: Is it 'tactical' deception? Animal Behaviour 64:185–195

Burkart JM, Hrdy SB, van Schaik CP (2009) Cooperative breeding and human cognitive evolution. Evolutionary Anthropology 18:175–186

Call J, Tomasello M (2008) Does the chimpanzee have a theory of mind? 30 years later. Trends in Cognitive Science 12:187–192

Chapais B (2010) The deep structure of human society: Primate origins and evolution. In: Kappeler PM, Silk JB (eds) *Mind the Gap: Tracing the Origins of Human Universals*. Springer, Heidelberg, pp 19–51

Church DM, Goodstadt L, Hillier LW, Zody MC, Goldstein S, She X, Bult CJ, Agarwala R, Cherry JL, DiCuccio M, Hlavina W, Kapustin Y, Meric P, Maglott D, Birtle Z, Marques AC, Graves T, Zhou S, Teague B, Potamousis K, Churas C, Place M, Herschleb J, Runnheim R, Forrest D, Amos-Landgraf J, Schwartz DC, Cheng Z, Lindblad-Toh K, Eichler EE, Ponting CP, The Mouse Genome Sequencing C (2009) Lineage-specific biology revealed by a finished genome assembly of the mouse. PLoS Biol 7:e1000112

Cirelli C, Bushey D, Hill S, Huber R, Kreber R, Ganetzky B, Tononi G (2005) Reduced sleep in *Drosophila* shaker mutants. Nature 434:1087–1092

Clayton NS, Dickinson A (1998) Episodic-like memory during cache recovery by scrub jays. Nature 395:272–274

Dickson BJ (2008) Wired for sex: The neurobiology of *Drosophila* mating decisions. Science 322:904–909

Enard W, Khaitovich P, Klose J, Zöllner S, Heissig F, Giavalisco P, Nieselt-Struwe K, Much-more E, Varki A, Ravid R, Doxiadis GM, Bontrop RE, Pääbo S (2002) Intra- and interspecific variation in primate gene expression patterns. Science 296:340–343

Evans PD, Gilbert SL, Mekel-Bobrov N, Vallender EJ, Anderson JR, Vaez-Azizi LM, Tishkoff SA, Hudson RR, Lahn BT (2005) Microcephalin, a gene regulating brain size, continues to evolve adaptively in humans. Science 309:1717–1720

Fichtel C, Kappeler PM (2010) Human universals and primate symplesiomorphies: Establishing the lemur baseline. In: Kappeler PM, Silk JB (eds) *Mind the Gap: Tracing the Origins of Human Universals*. Springer, Heidelberg, pp 395–426

Fehr E, Gächter S (2002) Altruistic punishment in humans. Nature 415:137–140

Fessler DMT, Gervais M (2010) From whence the captains of our lives: Ultimate and phylogenetic perspectives on emotions in humans and other primates. In: Kappeler PM, Silk JB (eds) *Mind the Gap: Tracing the Origins of Human Universals*. Springer, Heidelberg, pp. 261–280

Gat A (2010) Why war? Motivations for fighting in the human state of nature. In: Kappeler PM, Silk JB (eds) *Mind the Gap: Tracing the Origins of Human Universals*. Springer, Heidelberg, pp 197–220

Gintis H, Bowles S, Boyd R, Fehr E (2003) Explaining altruistic behavior in humans. Evolution and Human Behavior 24:153–172

Gurven M (2004) To give and to give not: The behavioral ecology of human food transfers. Behavioral and Brain Science 27:543–583

Hammock EAD, Young LJ (2005) Microsatellite instability generates diversity in brain and sociobehavioral traits. Science 308:1630–1634

Haygood R, Fedrigo O, Hanson B, Yokoyama K-D, Wray GA (2007) Promoter regions of many neural- and nutrition-related genes have experienced positive selection during human evolution. Nature Genetics 39:1140

Henke W, Tattersall I (2007) *Handbook of Paleoanthropology* (3 vols). Springer, Berlin

Hill K, Barton M, Hurtado AM (2009) The emergence of human uniqueness: Characters underlying behavioral modernity. Evolutionary Anthropology 18:187–200

Hrdy SB (2009) *Mothers and Others: The Evolutionary Origins of Mutual Understanding*. Harvard University Press, Cambridge MA

Kaplan H, Hill K, Lancaster J, Hurtado AM (2000) A theory of human life history evolution: Diet, intelligence, and longevity. Evolutionary Anthropology 9:156–185

Kappeler PM, Kraus C (2010) Levels and mechanisms of behavioural variability. In: Kappeler PM (ed) *Animal Behaviour: Evolution and Mechanisms*. Springer, Heidelberg, pp 655–684

Kappeler PM, Silk JS, Burkart JM, Schaik CP (2010) Primate behavior and human universals: Exploring the gap. In: Kappeler PM, Silk JB (eds) *Mind the Gap: Tracing the Origins of Human Universals*. Springer, Heidelberg, pp 3–15

Lovejoy CO (1981) The origin of man. Science 211:341–350

McElreath R (2010) The coevolution of genes, innovation, and culture in human evolution. In: Kappeler PM, Silk JB (eds) *Mind the Gap: Tracing the Origins of Human Universals*. Springer, Heidelberg, pp 451–474

McElreath R, Boyd R, Richerson PJ (2003) Shared norms and the evolution of ethnic markers. Current Anthropology 44:122–130

Marcus GF, Fisher SE (2003) FOXP2 in focus: What can genes tell us about speech and language? Trends in Cognitive Sciences 7:257–262

Markram H (2006) The blue brain project. Nature Reviews Neuroscience 7:153–160

Mikkelsen TS, Hillier LW, Eichler EE, Zody MC, Jaffe DB, Yang S-P, W. E, Hellmann I, Lindblad-Toh K, Altheide TK, Archidiacono N, Bork P, Butler J, Chang JL, Cheng Z, Chinwalla AT, deJong PJ, Delehaunty KD, Fronick CC, Fulton LL, Gilad Y, Glusman G, Gnerre S, Graves TA, Hayakawa T, Hayden KE, Huang X, Ji H, Kent WJ, King M-C, Kulbokas EJ, Lee MK, Liu G, Lopez-Otin C, Makova KD, Man O, Mardis ER, Mauceli E, Miner TL, Nash WE, Nelson JO, Pbo S, Patterson NJ, Pohl CS, Pollard KS, Prfer K, Puente XS, Reich D, Rocchi M, Rosenbloom K, Ruvolo M, Richter DJ, Schaffner SF, Smit AFA, Smith SM, Suyama M, Taylor J, Torrents D, Tuzun E, Varki A, Velasco G, Ventura M, Wallis JW, Wendl MC, Wilson RK, Lander ES, Waterston RH (2005) Initial sequence of the chimpanzee genome and comparison with the human genome. Nature 437:69–87

Orr HA (2005) The genetic basis of reproductive isolation: Insights from *Drosophila*. Proceedings of the National Academy of Sciences USA 102:6522–6526

Price TAR, Hodgson DJ, Lewis Z, Hurst GDD, Wedell N (2008) Selfish genetic elements promote polyandry in a fly. Science 322:1241–1243

Reppert SM, Weaver DR (2002) Coordination of circadian timing in mammals. Nature 418:935–941

Robinson GE, Fernald RD, Clayton DF (2008) Genes and social behavior. Science 322:896–900

Robson SL, van Schaik CP, Hawkes K (2006) The derived features of human life history. In: Hawkes K, Paine RL (eds) *The Evolution of Human Life History*. School of American Research Press, Santa Fe, pp 17–44

Rosso L, Keller L, Kaessmann H, Hammond RL (2008) Mating system and avpr1a promoter variation in primates. Biology Letters 4:375–378

Silk JB (2009) Nepotistic cooperation in non-human primate groups. Philosophical Transactions of the Royal Society B: Biological Sciences 364:3243–3254

Silk JB, Boyd R (2010) From grooming to giving blood: The origins of human altruism In: Kappeler PM, Silk JB (eds) *Mind the Gap: Tracing the Origins of Human Universals*. Springer, Heidelberg, pp 223–244

Smith EA, Borgerhoff Mulder M, Hill K (2001) Controversies in the evolutionary social sciences: A guide for the perplexed. Trends in Ecology and Evolution 16:128–135

Taborsky M, Brockman HJ (2010) Alternative reproductive tactics and life history phenotypes. In: Kappeler PM (ed) *Animal Behaviour: Evolution and Mechanisms*. Springer, Heidelberg, pp 537–586

Takahashi JS, Shimomura K, Kumar V (2008) Searching for genes underlying behavior: Lessons from circadian rhythms. Science 322:909–912

Tomasello M, Call J (1997) *Primate Cognition*. Oxford University Press, New York

Tomasello M, Moll H (2010) The gap is social: Human shared intentionality and culture. In: Kappeler PM, Silk JB (eds) *Mind the Gap: Tracing the Origins of Human Universals*. Springer, Heidelberg, pp 331–349

Tomasello M, Carpenter M, Call J, Behne T, Moll H (2005) Understanding and sharing intentions: The ontogeny and phylogeny of cultural cognition. Behavioral and Brain Sciences 28:675–735

Sachser N, Kaiser S (2010) The social modulation of behavioral development. In: Kappeler PM (ed) *Animal Behaviour: Evolution and Mechanisms*. Springer, Heidelberg, pp 505–536

van Schaik CP (2010) Social learning and culture in animals. In: Kappeler PM (ed) *Animal Behaviour: Evolution and Mechanisms*. Springer, Heidelberg, pp 623–654 van Schaik CP, Burkart JM (2010) Mind the gap: Cooperative breeding and the evolution of our unique features. In: Kappeler PM, Silk JB (eds) *Mind the Gap: Tracing the Origins of Human Universals*. Springer, Heidelberg, pp 477–496

Vargha-Khadem F, Gadian DG, Copp A, Mishkin M (2005) FOXP2 and the neuroanatomy of speech and language. Nature Reviews Neuroscience 6:131–138

Wang Y-q, Su B (2004) Molecular evolution of microcephalin, a gene determining human brain size. Human Molecular Genetics 13:1131–1137

Whiten A (2010) Ape behavior and the origins of human culture. In: Kappeler PM, Silk JB (eds) *Mind the Gap: Tracing the Origins of Human Universals*. Springer, Heidelberg, pp 429–450

Whitfield CW, Cziko A-M, Robinson GE (2003) Gene expression profiles in the brain predict behavior in individual honey bees. Science 302:296–299

Chapter 2
Our Children: Parental Decisions — How Much to Invest in Your Offspring

Mary K. Shenk

Abstract Reproduction is the most fundamental of evolutionary behaviors, yet human parents face especially complex tradeoffs when deciding how many children to have and how much to invest in each of them. This chapter reviews parental investment theory, including both the key concepts and some important questions to which they have been applied in humans. Written primarily from the perspective of human behavioral ecology, this chapter also discusses how evolutionary social scientists have approached cross-cultural variation in parenting behavior. The chapter begins with an overview of life history theory and the concept of reproductive tradeoffs, focusing especially on the tradeoffs between current vs. future reproduction and quantity vs. quality of offspring. Discussing the critical question of who invests in offspring, I next compare motivations for investment between mothers and fathers, and explore the roles of many types of kin in investment, while considering whether humans can be viewed as cooperative breeders. I then explore the role of parent–offspring conflict and sibling conflict in parental investment and inheritance systems, followed by an exploration of sex biases in investment, including the Trivers–Willard effect local resource competition, and local resource enhancement. In conclusion, I argue that parental investment has been one of the most active areas of enquiry among evolutionary researchers over the last twenty years, and is likely to remain one of the mainstays of the field during the coming decades.

2.1 Introduction to Parental Investment Theory

How many children should parents have, and how much time and resources should they invest in each of them? These are the two central questions at the heart of evolutionary approaches to parental investment. Each question implies a tradeoff

Mary K. Shenk
Department of Anthropology, University of Missouri, 107 Swallow Hall, Columbia, MO 65211-1440, USA, e-mail: shenkm@missouri.edu

U.J. Frey et al. (eds.), *Essential Building Blocks of Human Nature*, The Frontiers Collection, DOI 10.1007/978-3-642-13968-0_2, © Springer-Verlag Berlin Heidelberg 2011

— the first between current and future reproduction, and the second between offspring quantity and offspring quality. The study of parental investment centers on the ways in which different ecological and cultural circumstances change the costs and benefits of investment, causing people to arrive at different answers to these two questions.

In his seminal 1972 article, Trivers defined parental investment as "any investment by the parent in an individual offspring that increases the offspring's chance of surviving [...] at the cost of the parent's ability to invest in other offspring" (Trivers 1972, p. 139). Other perspectives have expanded the concept to include many types of behavior which increase an offspring's fitness while reducing the resources parents have to invest in other offspring, self-maintenance, future reproduction, or aid to other kin; the concept has also been extended to kin other than parents (Clutton-Brock 1991; Hamilton 1964).

Thus defined, parental investment includes most types of direct care, including gestation, lactation, food provisioning, protection, and the education/training of offspring. Complex human social systems also create opportunities for additional forms of parental investment not typically thought of as parental care, such as arranging marriages, transferring social connections, and endowing offspring with wealth (Alexander 1990; Trivers 1972). The benefits of parental investment to offspring can also take many forms, including effects on survival, growth, health, immune function, and social status, which ultimately affect fitness, the number of offspring (or other close kin discounted by their degree of relatedness) surviving in future generations.

The benefits provided to offspring, however, come at costs to parents. Time spent rearing offspring can lead to lost mating opportunities or a delay before parents can have another offspring, while resources spent rearing offspring can reduce the resources left for other offspring or impede a parent's ability to repair and maintain their own body. Parental investment theory predicts that parents have been shaped by natural selection to maximize the difference between the benefits and the costs of parental investment, which they do by making tradeoffs in offspring number and/or the care that each offspring receives.

This chapter is written primarily from the perspective of human behavioral ecology (HBE), or human evolutionary ecology. Deriving from the study of animal behavior, HBE and HEE attempt to explain human behaviors as adaptive solutions to the competing demands of growth and development, knowledge and status acquisition, mate acquisition, reproduction, and parental care (Smith & Winterhalder 1992). HBE and HEE also aim to understand variation in human behavior both across and within human cultures.

This paper aims not only to introduce parental investment theory, but additionally to provide an overview of some of the many important questions and topics to which it has been applied in humans. I will begin with an overview of the classic theoretical insights that originated the study of parental investment, including life history theory and the concept of reproductive tradeoffs. I will then discuss the important issue of who cares for children, exploring the roles of parents and other relatives. Next I will discuss two key issues involving inequality in investment: parent–offspring and sibling conflicts, and sex biases and the Trivers–Willard effect. In each section I will explore some of the primary themes in the literature, introducing theory along with examples of relevant research.

2.2 Life History Theory and Tradeoffs

The evolutionary ecology perspective on reproduction is centered on the idea that human fertility and mortality decisions follow the patterns expected under life history theory, a branch of evolutionary biology that focuses on the relationships between growth, reproduction, and survival across the life cycle (see, e.g., Borgerhoff Mulder 1992; Chisholm 1993; Clarke & Low 2001; Hill 1993; Hill & Kaplan 1999; Low 1998; Mace 2000). Insights and predictions from life history theory bear directly on key questions about reproduction, including the timing of births, the number of offspring born, how much parents invest in children, and the relationship between mortality and fertility. Evolutionary researchers have applied this perspective to cultures in varying ecological circumstances and have found strong evidence for its utility in understanding and predicting both individual reproductive behavior and human population processes.

According to life history theory, an organism's lifetime is characterized by tradeoffs between investment in somatic effort and reproductive effort (Borgerhoff Mulder 1992; Chisholm 1993). Somatic effort is investment in development, growth, and maintenance of the body, while reproductive effort is investment in any aspect of reproduction; the latter is usually subdivided into mating effort and parenting effort (Alexander & Borgia 1979). Tradeoffs exist because any time or energy invested in one type of effort reduces the amount of time or energy that can be invested in the other. Lessels (1991) argues that many major life history tradeoffs can be subsumed into two categories relevant to reproduction:

- the tradeoff between current and future reproduction,
- the tradeoff between offspring quantity and offspring quality.

2.2.1 Current vs. Future Reproduction

Perhaps the most fundamental reproductive tradeoff is that between beginning reproduction and continuing investment in somatic maintenance, growth, development, and/or the acquisition of knowledge, status, or mates (Chisholm 1993). Generally, when mortality risks are high, organisms do better by maturing faster and reproducing early in life. This is because the longer an organism delays reproduction, the more it increases its probability of dying before it gets a chance to reproduce. When mortality is low, however, organisms can afford to delay reproduction and prolong somatic investment in the hope of greater reproductive success later in life.

An important concept in the study of this tradeoff is reproductive value (RV). Defined by Fisher (1958), RV is the number of offspring an individual of a given age can expect to produce in the remaining years of its life, adjusted by the probability of surviving each of those years. At any given age, reproductive value can be partitioned into current reproductive value and residual reproductive value or RRV (Williams 1966). Increasing or continuing investment in current offspring reduces the resources available to future offspring, and thus RRV, whereas decreasing or withdrawing investment from current offspring increases the resources available to future offspring, increasing RRV.

For women, RRV increases sharply until around age twenty, at which point it begins to fall off quickly, reaching zero at menopause around age 50; for men reproductive value increases more slowly until the early 20s, then tails off more gradually, reaching zero around age 70 (see, e.g., Hamilton 1966). Very early reproduction can entail significant costs in terms of growth and development as well as the ability to invest later in life. For example, Nigerian women who had their first births as teenagers showed impaired growth compared to women who had later first births (Harrison et al. 1985). Alternatively, delaying reproduction too long can also entail costs in terms of both fertility (the number of children born) and fecundability (the ability to conceive). For example, highly educated women in modern developed nations often delay childbearing so long that they have significant problems with conception and childbirth, reducing fertility levels and sparking demand for medical technologies to treat infertility (Kaplan et al. 2002).

Because they do not undergo menopause, men can afford to delay reproduction longer than women without significantly reducing lifetime reproductive success. Moreover, in low mortality settings with large wealth differentials, delaying reproduction can increase male reproductive success if that time is used to increase wealth or status (Miller 2000). Females may also get some benefits from delay in societies where they also compete for wealth or status, but women face more serious fertility consequences, because delaying reproduction does not delay menopause or change the steep decline in RRV with age.

2.2.2 Quantity vs. Quality

The second fundamental tradeoff is between parental investment (the resources and care expended on each offspring) and fertility (the number of offspring born and reared) (Trivers 1972). High levels of parental investment in existing offspring necessarily require lower fertility, while low levels of parental investment in offspring allow for higher fertility.

MacArthur and Wilson (1967) defined two general strategies organisms can take in negotiating this tradeoff. An r-selected strategy occurs in unpredictable environments with high mortality rates, where it is important to reproduce quickly or risk not being able to reproduce at all; r-selected species have early ages at maturation, high fertility, and low levels of parental investment. In more stable populations which are nearer the carrying capacity (K) of an environment, however, within-species competition for resources increases. In these circumstances, natural selection favors organisms that have later ages at maturation and fewer offspring, but invest more heavily in each. Primates as an order are K-selected organisms, known for their high levels of parental investment. Most species of Old World monkeys and apes (our closest relatives) bear one offspring at a time, engage in a lengthy period of lactation, and have relatively long interbirth intervals lasting from one to eight years (Alvarez 2000). Humans fit much of this pattern. We usually only give birth to a single offspring at a time, periods of breastfeeding last from 2–4 years in traditional societies, and children are not self-sufficient or productive foragers until well into their teens or even later (see, e.g., Borgerhoff Mulder 1992; Hill & Hurtado 1996; Kaplan 1996; Lee 1979). While the terms r-selection and K-selection are usually used to designate differences between species, they are sometimes applied to differences within species. For example, Wilson and Daly (1997) found that women in Chicago neighborhoods with high mortality rates had both higher fertility rates and earlier ages at first birth (and thus relatively more r-selective reproductive strategies) than women in neighborhoods with lower mortality rates.

Interbirth Intervals

In humans as in other species, the amount of time between the birth of one offspring and the next is one measure of parental investment. Longer interbirth intervals are generally associated with higher levels of parental investment or a more quality-oriented strategy, while shorter birth intervals are associated with a low parental investment and high fertility strategy (Borgerhoff Mulder 1992). Humans are unusual, however, in that we customarily have several dependent offspring at one time, meaning that short interbirth intervals may affect not only the most recent child but older children as well.

Humans in many foraging societies have moderate to very long interbirth intervals sustained by long periods of lactation and lactational amenorrhea (see, e.g., Lee 1979). For example, Lee finds that among the !Kung of the Kalahari Desert, interbirth intervals of nearly 4 years were sustained by women who faced the high workloads of active foraging, whereas interbirth intervals shortened by nearly a year when women settled on cattle stations and led more sedentary lives. In many cases the length of these birth intervals is positively associated with child well-being. For example, Blurton Jones (1986, 1987) found that, except for the first, shorter interbirth intervals were linked to higher rates of child mortality among the !Kung.

The same basic tradeoff holds among many non-foraging cultures as well. For example, Gibson and Mace (2006) found that an increase in birth rates caused by the introduction of wells into villages in Ethiopia was associated with decreased child nutritional status and higher mortality when siblings were very closely spaced. In two large comparative studies of 26 and 39 developing nations, Hobcraft, McDonald, and Rutstein (1985) found that birth intervals shorter than 2 years were associated with significantly higher risks of infant and child mortality than were longer birth intervals. Similar effects are also found in the developed world. Among women from the United Arab Emirates, short inter-pregnancy intervals were associated with preterm births, a significant risk factor for child mortality and developmental complications (Al-Jasmi et al. 2002).

Mortality and Risk

Quantity/quality tradeoffs are heavily influenced by mortality rates. In general, when mortality rates are high, natural selection should favor a strategy of higher fertility and lower parental investment, while when mortality rates are low, natural selection should favor a strategy of lower fertility and higher parental investment (Gasser et al. 2000). Chisholm (1993) argues that individuals who experience high mortality environments as children are more likely to reproduce early and often and invest less per child, while those who experience lower mortality environments are more likely to delay reproduction and pursue a strategy of lower fertility and higher parental investment (see also Draper & Harpending 1982).

When discussing on the effects of mortality on human fertility, some authors emphasize the difference between care-dependent and care-independent forms of mortality or, more generally, risk (see, e.g., Quinlan 2007). Care-dependent, or intrinsic, forms of risk can be reduced by the actions of parents or other caretakers. On the other hand, extrinsic or care-independent types of risk cannot easily be ameliorated by caretakers. High levels of intrinsic risk are predicted to lead to increased parental investment, while high levels of extrinsic risk should lead to lower levels of parental investment. Quinlan and Quinlan (2007) use parental investment data from the *Standard Cross-Cultural Sample* to argue that unresponsive parenting practices may be adaptive in high-risk environments, such as those associated with high levels of extramarital sex, aggression, theft, and witchcraft. Using the same data, Quinlan (2007) finds that the level of maternal care is inversely correlated with such extrinsic risks as famine, warfare, and high levels of pathogen stress.

Family Size

Some authors have attempted to demonstrate a quantity–quality tradeoff in particular human populations directly by comparing fertility and child survival to determine whether an intermediate number of children appears to be optimal. Blurton Jones (1987) finds that, among the !Kung foragers of the Kalahari desert, intermediate numbers of children are optimal with respect to child survival. Hagen et al. (2006) find that, among the forager–horticulturalist Shuar, children living in households with a higher consumer-to-producer ratio were shorter and had lower nutritional status. Shorter and lighter children are at risk of increased mortality and decreased fertility rates, suggesting that an intermediate number of offspring might be most likely to optimize long-term fitness in this population. And Strassmann and Gillespie (2002) find that large numbers of siblings are associated with reduced child survival among the agricultural Dogon of Mali, while an intermediate number of children is ideal.

2.3 Who Invests: Mothers, Fathers, Grandmothers, and Others

Investment in offspring can be provided by the mother and father, but also by other relatives and group members. The question of who cares for children has been the subject of a great deal of research in recent years, with attention being paid especially to two hallmarks of human investment behavior:

- non-maternal care,
- variations in patterns of investment between societies.

2.3.1 Mothers and Fathers

Drawing on seminal work by Bateman (1948) and others, Trivers (1972) argued that, within species, the sex which invests more heavily in offspring will become a limiting resource for the other sex. Among mammals, the higher investing sex is most often females since they bear a much larger burden of obligatory parental investment (Clutton-Brock 1989). Among humans, for example, the minimum amount of male parental investment is the sperm necessary for conception. For women, in contrast, the minimum amount of parental investment required is the much larger egg cell, 40 weeks of gestation, and a period of lactation that in traditional human societies can last for several years. This suggests that, in general, humans should follow the stereotypical pattern of high female parental investment and male competition for choosy females.

Yet levels of parental investment vary significantly both between and within species, and humans are a highly variable species in terms of male and female contributions to parental investment (see, e.g., Brown et al. 2009; Hill & Kaplan 1999; Voland 1998). In fact, Trivers' model predicts that, in species with high paternal investment, females should compete for access to males since it is male investment which may be the limiting factor on female reproductive success (Trivers 1985). Moreover, Brown et al. (2009) argue that a careful reading of Bateman's principles and cross-cultural comparison of human mating and population structures shows that, since human cultures vary in characteristics such as population density and sex ratio, Bateman gradients should differ between human populations. The authors find that, while men on average have greater variance in reproductive success than women, suggesting that men should compete for women who invest more, this is not always true. Moreover, though monogamous societies generally have smaller differences between male and female RS than polygynous ones, indicating a greater motivation for male parental investment in monogamous groups, there is tremendous variation across both polygynous and monogamous societies.

One thing that varies little, however, is that mothers are the most important caretakers of young children in most human societies. For example, in a recent review article entitled *Who keeps children alive?*, Sear and Mace (2008) found that mothers have a virtually universal — and often profound — positive effect on child survival, while maternal death is often associated with extraordinarily high rates of mortality for young children. Since maternal care is so pervasive, it shows less cross-cultural variability than care by other kin or group members. Yet how and how much mothers invest does vary between societies and individuals, often in ways that are consistent with levels of resources, risk, and opportunity costs (Hrdy 1999; Voland 1998). In general, mothers who are healthy, have abundant resources, face lower opportunity costs, have fewer helpers, or expect high returns on parental investment invest more per child, while mothers who are less healthy, lack resources, face greater opportunity costs, have more helpers, or expect low returns on parental investment may choose to limit investment per child or terminate investment in particular children entirely (Hrdy 1992; Kaplan 1996; Trivers 1974; Trivers & Willard 1973).

While human mothers perform the bulk of care for infants and children, humans are unusual among higher primates (monkeys and apes) for the high amount of investment in offspring by fathers (Marlowe 2000). Investment by fathers has long been touted as one of the hallmarks of the human species, and the importance of fathers in provisioning and/or protecting mates and children has figured prominently in many conceptions of human evolutionary history (see, e.g., Lee & DeVore 1968). Both classic and recent models suggest that paternal aid in the provisioning and care of children has helped to fund the high reproductive rates of humans as compared to other primate species (Kaplan et al. 2000; Marlowe 2000). Some also argue that paternal investment is a key to the development of important human cognitive and social abilities (Flinn et al. 2007; Geary 2000; Hrdy 2009). While there is disagreement about just how important paternal investment has been in our evolutionary history, most researchers agree that our capacity for high levels of paternal investment is a key evolved feature of human behavior.

Among the highest levels of direct paternal investment are found among the Aka peoples of the Central African Republic. In these groups, fathers spend a great deal of time holding infants and playing with young children, in some cases nearly rivaling the time spent by mothers (Hewlett 1991). The same pattern appears to be found in many other forager groups as well (Marlowe 2000). Fathers also appear to be highly important in monogamous societies with high levels of parental investment, such as modern nations in Europe and North America (Harrell 1997; Marlowe 2000). In other cases — most frequently among horticulturalists — the role of fathers is more flexible, with fathers being key players in some families but not others (Leonetti et al. 2007; Marlowe 2000).

There are other cases in which fathers appear to be less important or even uninvolved. Sear and Mace (2008) find, for example, that in over half of the societies for which they had good data, fathers had no effect on child survival. In contrast, Scelza (2010) suggests that the limited importance of fathers found in some societies may be an artifact of the measures commonly used as indicators of investment, including child mortality and child nutritional status, which may underestimate the importance of fathers if their contributions come later in the life cycle, during adolescence or adulthood. For example, Scelza (2010) finds that adolescent Martu aborigine boys with an absent father or father figure have delayed age at initiation and consequently delayed age at first birth.

2.3.2 Grandmothers and Others

Humans show evidence of a great deal of alloparenting, or care given by people other than the parents, especially in traditional societies (Hrdy 2009; Mace & Sear 2005). The complexity of human social systems also allows for a great deal of variation between societies — far more variation than is common among other primates. Recent work has emphasized the importance of grandparents, aunts and uncles, siblings, and occasionally others in keeping children alive and improving their prospects for finding mates and reproducing.

Most famous of the perspectives on the contribution of non-parental kin has been research on the importance of grandmothers. Based on their work among the Hadza foragers of Tanzania, Hawkes and colleagues (1997) developed the 'grandmother hypothesis' as an explanation for the evolution of menopause in humans. Menopause is a very rare event among animals, and its origins are hotly debated. Hawkes et al. argued that humans evolved the ability to have multiple dependent offspring at one time because the additional work it took to provision and care for simultaneous children was subsidized by post-menopausal women working alongside their daughters or daughters-in-law to feed and care for their grandchildren. The crux of this model is the suggestion that older women get a better reproductive payoff by investing in grandchildren than they would get from continuing personal reproduction into an increasingly feeble old age.

The grandmother hypothesis has been both influential and controversial (see, e.g., Voland et al. 2005). An alternative perspective is provided by Williams (1957) who suggested that, given the very long dependency periods of human children, it was just as likely that women ceased reproducing in order to successfully rear their own offspring to adulthood. Others have argued that menopause is better understood as a result of the biology of human ovaries, which are optimized to maintain regular cycles at young ages (see, e.g., Wood et al. 2001). Despite these critiques, a great deal of work has been done by numerous researchers to test the implications of the grandmother hypothesis in societies around the world (Voland et al. 2005). These results can best be described by revisiting the review article by Sear and Mace (2008), who find substantial cross-cultural support for the importance of grandmothers (though their impact is not always positive), but also for the importance of other relatives (including fathers, sisters, aunts, and occasionally grandfathers) as well as a great deal of variation between societies in terms of which relatives helped or hurt children's survival. This work suggests that, while grandmothers can be very important, they are not universally or uniquely so.

Various authors have considered the importance of siblings to parental investment. Kramer and Boone (2002) argue that high fertility among intensive agriculturalists is underwritten by the labor of children on the farm. Using data on Maya children in rural Mexico, they show that, while children may not be self-sufficient at early ages, the work they do reduces the workload of their parents, freeing up time and calories that can be spent on subsequent reproduction. Kramer (2005) further argues that partial self-provisioning by children is common in traditional societies following many subsistence patterns (hunter–gatherer, pastoralist, agriculturalist) and contributes to the human capacity for high reproductive rates. Other ways that siblings can be important are considered in Sect. 2.5.

Several studies have also examined the effects of investment by aunts and uncles. In a study of US college students, Gaulin et al. (1997) find that, while aunts generally invest more than uncles, maternal relatives invest more than paternal relatives; they interpret this outcome in terms of the greater risk of paternity uncertainty on the part of patrilateral relatives. Similar results were found by Pashos and McBurney (2008). Shenk (2005) found that the education and wealth of aunts and uncles had a positive effect on children's education and income, but that the numbers of aunts and uncles had a negative effect. Sear and Mace (2008) also report uneven results for the effects of aunts and uncles on child survival, with positive and negative effects of both aunts and uncles in several historical or traditional populations.

Human societies are also unusual in that they regularly engage in the adoption and fostering of children. These practices are especially common in traditional societies in Africa and Oceania, but are nearly universal in some form among human groups (see, e.g., Harrell 1997). Some authors have argued that these customs, especially where they are common, are probably related to the need to adjust the number or genders of dependents between households, as either too few or too many may cause resource stress and lead to inadequate provisioning or care of children (Harrell 1997; Silk 1980; Turke 1988). As predicted by the principle of kin selection (Hamilton 1963), in traditional societies adoption and fostering are most common among close kin.

2.3.3 Are Humans Cooperative Breeders?

The importance of relatives to the care of children in so many societies, combined with cross-cultural diversity in terms of whom the key relatives are, has caused several authors to draw the conclusion that humans should be characterized as cooperative breeders (Hrdy 2009; Kramer 2005; Mace & Sear 2005; Sear & Mace 2008). Kramer (2005) and Sear and Mace (2005) argue that our history as cooperative breeders is the best explanation for the high rates of fertility in humans, as compared to our close primate relatives. Hrdy (2009) further contends that cooperative investment has been a key in the evolution of our psychological adaptations for empathy and cooperation, which are unique among animals.

While many authors have come to question Hawkes et al.'s (1997) perspective on the primacy of grandmaternal investment, there is increasing agreement that cooperative breeding strategies (of which grandmothers are in many cases a key part) are in fact related to the evolution of menopause (Sear & Mace 2005). Early attempts to model or predict menopause suggested that the inclusive fitness benefits of mothering or grandmothering were not sufficient to offset the potential benefits of continuing to reproduce (Hill & Hurtado 1991, 1996; Rogers 1993). Cant and Johnstone (2008), however, argue that if one takes the costs of reproductive competition into account, one reaches a different conclusion. They argue that, in the context of female dispersal, which likely characterized early human societies, older women will compete with younger, immigrant women who have a competitive advantage, because they are insensitive to the reproductive costs of older females and have less to gain from cooperation. In these circumstances, the benefits of continued reproduction may become low enough that there would be selection for reproductive cessation. If post-menopausal women are able to focus investment on older children and grandchildren, this would only strengthen the benefits of cessation.

2.4 Parent–Offspring and Sibling Conflicts

Reproducing and investing in offspring is enormously costly in terms of time and other resources. Individuals are limited in the amount of energy they can devote to producing and raising young, and such expenditures may be detrimental to the investor's own condition, survival, and future reproductive output. However, investment is typically beneficial to the offspring themselves, enhancing their condition, survival, and reproductive success. These differences in the perspectives of parents and offspring may come into conflict at times, leading to what Trivers (1974) referred to as parent–offspring conflict. Such conflicts can begin very early in pregnancy, continue throughout infancy and childhood, and in some circumstances extend through the marriage of a child or the division of parental resources among children. A closely related type of conflict occurs between siblings when they compete with each other for access to resources or mates (see, e.g., Trivers 1974, 1985). As with parent–offspring conflict, sibling conflict often begins at young ages, but in humans can extend throughout the lifespan. This section will cover two key types of parent–offspring and sibling conflicts; there are many others which I do not have space for here.

2.4.1 Infanticide and Neglect

Human parents often have to make difficult decisions about how to divide resources between existing older offspring, new infants, and potential future offspring. One of the most widely studied examples of how parents manage this tradeoff is infanticide, a phenomenon which exists in some form in many animals, including many species of primates and most human societies (see, e.g., Daly & Wilson 1988; Hrdy 1992). In a survey of the infanticide literature, Daly and Wilson (1988) found that most human infanticide is attributable to one of three circumstances: poor offspring quality, lack of parental resources, or lack of certainty regarding paternity. Each of these has clear evolutionary implications. Research on the non-lethal neglect of children often shows similar patterns.

   Lack of parental resources commonly affects fertility and parental investment decisions in two ways. First, younger women may use infanticide or neglect to delay childbearing or investment until they are in more favorable circumstances. Second, older or poorer women may use investment or neglect as a means of protecting their investment in existing children, as large numbers of children or closer birth spacing can increase the level of competition between siblings and put the investment already made in older offspring at risk. For example, both types of motivations can be found for abortion in modern Western societies. Young, unmarried women are often the most common abortion patients, especially when they do not have a supportive partner or they feel as though having a child would put their education or job opportunities at risk (Finer et al. 2005; Hill & Low 1991; Lycett & Dunbar 1999). Women in their 30s or older, on the other hand, usually cite poverty or responsibilities to older children as their primary reasons for seeking an abortion (Finer et al. 2005). Many authors have found that abortion rates among single women drop as they become older, and argue that, as the likelihood of finding a better future circumstance for childbearing becomes lower, current reproductive effort becomes more valuable than future effort (Hill & Low 1991; Lycett & Dunbar 1999; Tullberg & Lummaa 2001).

Children with physical deformities or mental handicaps are at high risk of abuse, neglect, and infanticide cross-culturally (Daly & Wilson 1984). If parents are poor enough, even more subtle signs of low infant quality such as listlessness or failure to thrive may be cues that cause parents to limit investment. In the poor shantytowns of Northeast Brazil, for instance, Scheper-Hughes (1992) describes maternal decision-making in the context of extreme poverty and very high infant mortality rates. Instead of investing more in children who appeared weak or sick, mothers often reduced investment and instead targeted care and resources towards infants who showed higher energy levels and greater frequency of crying — infants they interpreted as more likely to survive and/or thrive in their challenging environment.

Twins are also at higher risk of infanticide cross-culturally (Ball & Hill 1996). Twins are problematic for two reasons. First, twinning is associated with prematurity, low birth weight, and occasionally other developmental anomalies, all of which may put children at risk of higher mortality as well as physical and mental problems at later ages (Ball & Hill 1996). Second, in many traditional cultures it is difficult for mothers or their kin to provide sufficient breast milk or care for two infants at the same time, thus imposing stress on children and families which can lead to problematic outcomes. For example, in 18th and 19th century Germany, Gabler and Voland (1994) found that both twins and their mothers suffered higher mortality rates than women who had single children, and that women who had twins paid a fitness cost in terms of lower numbers of surviving grandchildren. While Sear et al. (2001) found that in Gambia mothers of twins had more surviving children, this was despite the much higher rates of stillbirth and neonatal mortality among twins as compared to singletons.

Paternity certainty is an important evolutionary consideration since in many societies men may limit or terminate investment in offspring they believe are not their own. In a cross-cultural analysis of infanticide using the 60 societies in the *HRAF Probability Sample*, Daly and Wilson (1984, 1988) report non-paternity as a reason given for infanticide in 20 societies with specific concerns ranging from adulterous conceptions (the most common), to non-tribal sires, to the fact that the children were from a woman's first marriage. Furthermore, Fuster (1984) finds that mortality rates were much higher for illegitimate than for legitimate infants in late 19th and early 20th century Galicia, and were especially high among children not recognized by either the mother or the father.

2.4.2 Differential Investment and Inheritance

Parent–offspring and sibling conflicts are especially common in societies where wealth, often in the form of animals or land, is inherited. In such circumstances siblings of one or both genders directly compete for familial resources which are limited in size and may not be easily divisible. A certain number of cows or amount of land, for instance, usually needs to remain intact in order to support a family. In some situations these kinds of constraints may lead parents to adopt restricted forms of inheritance such as primogeniture, where the oldest child (usually the oldest son) inherits most of the family property, or ultimogeniture, where the youngest son or daughter inherits most of the family property.

Among the Gabbra pastoralists of East Africa, Mace (1996) found that later-born sons married at older ages, were given smaller herds on marriage, and often had lower fertility than their older brothers. She argued that this was likely a deliberate strategy on the part of parents to make sure that some sons would be successful; alternatively, it could be viewed as a the result of competition among siblings which older sons are better positioned to win. Low (1991) found a similar effect in 19th century Sweden, in which the presence of older brothers reduced the fertility of younger brothers. Voland and Dunbar (1995) found similarly that, among 18th and 19th century land-owning farmers in the Krummhörn region of Germany, later-born sons and daughters had higher infant mortality rates than their older same-sex siblings; these relationships did not hold true among landless laborers in the same area, suggesting that sibling competition may be stronger in groups with heritable wealth. Celibacy was also a common practice in some agricultural societies of Europe and Asia where wealth was based on land ownership. Many families only allowed one son and one daughter to marry in each generation, while their siblings stayed home to help with the farm or migrated to find work elsewhere (Boone 1986; Hajnal 1965; Deady et al. 2006).

Such tradeoffs are also present, if not more exaggerated, in modern industrial environments. For example, Kaplan (1996) argues that in modern wage-labor economies with education-based job markets, fertility has the potential to reach very low levels because the perceived payoffs to parental investment do not diminish until very high levels. He also argues that this calculus is stronger for more highly-educated people since they are more efficient at investing in education for their own children, leading to very high levels of investment as well as an inverse relationship between wealth and fertility. Similarly, Lawson and Mace (2009) find that, in contemporary Britain, children from smaller families received more investment per child than children from larger families, even when the effects of wealth were adjusted for, and, moreover, that the effects of the tradeoffs were more negative for later-born siblings. They also found that, as parents became wealthier and better educated, the strength of the tradeoffs increased.

2.5 Sons vs. Daughters: Sex Biases in Parental Investment

Interest in the evolutionary relationship between parental investment and offspring sex began with Fisher (1930), who argued that, since sons and daughters receive equal genetic contributions from both parents, investing in one sex will, on average, yield the same effect on parental fitness as investing in the other. An interest in sex-biased investment began when Hamilton (1967) suggested that, when siblings of one sex compete with each other for mates, parents may benefit by producing more of the opposite sex. Perhaps the best-known perspective on sex bias in parental investment comes from the 1973 Science article by Trivers and Willard, who argued that parents should bias investment towards the sex of offspring with the greatest potential for reproduction. In situations where the variance in reproductive success is higher for males than for females (which is usually the case in mammals), their model predicts that parents in good condition should invest more heavily in sons to take advantage of their higher potential RS, while parents in poor condition should invest more heavily in daughters because they are more assured of reproducing.

Predictions from these models have been tested many times in human populations. Human cultures, however, vary in multiple ways that change the payoffs to parental investment to either favor or discourage gender biases in investment. We thus see a wide variety of gender-biased investment strategies in response to varying ecological and social conditions.

2.5.1 The Trivers–Willard Effect

There are many cultures in which parents have been shown to systematically bias investment towards one sex or the other based on parental characteristics such as health, wealth, or social status. Sex-based investment can take many forms, from alterations of the sex ratio itself through infanticide or abortion, to mild or extreme forms of neglect, discrimination or favoritism, to investing different types of resources or employing different strategies in raising and marrying sons vs. daughters.

Evidence of son-biased investment is typically found among high status families or social groups. For example, daughters in elite families in traditional North India and China were often subject to infanticide because their marriage prospects were limited by rules of hypergyny dictating that women marry men of the same or a higher social rank. Daughters of high-ranking families faced a circumscribed marriage market because there were few places for them to marry upwardly, whereas sons of elite families had good marriage prospects among lower-status women and in some cases were able to take multiple wives and/or concubines (Dickemann 1979). Similarly, several studies find that high-status fathers have more sons than average in modern environments (see, e.g., Cameron & Dalerum 2009; Hopcroft 2005). Hopcroft (2005) shows that sons of high-status fathers achieve more education than their sisters, while daughters of low-status fathers achieve more education than their brothers. Cameron and Dalerum (2009) argue that the high proportion of sons born to male billionaires compared to the general public is an adaptive strategy because the same population also leaves more grandchildren through sons than daughters.

In contrast, other groups show systematic evidence of daughter bias. For example, among the polygynous, pastoralist Mukogodo of Kenya, daughters receive more frequent breastfeeding and better medical care than sons because they have better marital prospects than their brothers, who are not sought after as husbands due to their low social position (Cronk 1991). Mukogodo men have lower rates of polygyny and lower reproductive success than Mukogodo women, making investment in daughters an adaptive strategy. Boone (1986) finds that, in medieval Portugal, daughters of the lower nobility had more children than their brothers, causing a shift in the investment of parental resources towards dowering daughters and helping to ensure that they married into high-status families. Finally, in their work on Hungarian Gypsies, Bereczkei and Dunbar (1997) show evidence for a female-biased sex ratio, higher investment in daughters, and a greater number of grandchildren through daughters than sons among urban Gypsies. Gypsies are at the lower end of the Hungarian socioeconomic spectrum, and girls have a much better chance to marry upwards than do boys. In response to this, Gypsy mothers breastfeed their daughters longer, are more likely to terminate a pregnancy after a daughter than a son, and pay for daughters to continue longer in school than sons.

2.5.2 Local Resource Competition and Enhancement

Two special cases of sex-biased investment take place when children of one sex either compete or cooperate with each other — or with their parents — in terms of subsistence or reproduction. Local resource competition occurs when children of one sex compete with each other or with their parents for the same types of resources, prompting parents to limit investment towards the competing sex in favor of the non-competing sex (Clarke 1978; Silk 1983; van Schaik & Hrdy 1991). Competition can occur over subsistence resources like food, heritable resources like land, or access to mates or mating opportunities. For example, Voland et al. (1997) compared six populations in early 19th century Germany and found that infant mortality rates for daughters were higher in areas where populations were increasing, while infant mortality rates for sons were higher in areas where populations were stable. The authors argue that sons could start new farms in unoccupied land, but once land ownership was saturated, sons competed with each other for land. In early modern Portugal, younger sons of noble families did not inherit the family estate but had to work for status and wealth in the army or trade. Boone (1986) found that, among the high nobility, younger sons had much higher mortality and lower fertility rates than older, inheriting sons. He also found that this difference became more pronounced as Portugal's land became more saturated and estates became fixed.

Local resource enhancement occurs when one sex aids parents in reproduction (Emlen et al. 1986; Gowaty & Lennartz 1985) or offspring of one sex enhance the mating success of parents or siblings (Sieff 1990). In such circumstances, parents will be motivated to invest relatively more in the enhancing sex. For example, Turke (1988) reported that, on the Micronesian atoll of Ifaluk, women with first-born daughters had on average 2 additional offspring compared to women with first-born sons; this effect was significantly enhanced when comparing women whose two first-born children were daughters (who averaged 8.9 surviving offspring) as opposed to women whose two first-born children were sons (who averaged 5.1 surviving offspring). Bereczkei and Dunbar (2002) found a similar trend among Hungarian Gypsies, where mothers with first-born daughters had shorter birth intervals and longer reproductive careers than mothers with first-born sons, leading to an average of one additional surviving offspring. This phenomenon, which was first described in birds, is often called 'helping-at-the-nest' (Emlen et al. 1986).

2.5.3 Marriage Payments as Sex-Biased Parental Investment

Marriage payments are common in human societies. Though there is variation in terms of the content of the payment and who pays whom, virtually all forms of marriage payments are sex-biased: they are given to one sex but not the other, or given to sons and daughters in different amounts (Goody & Tambiah 1973; Harrell 1997). Additionally, marriage payments can often be viewed as forms of parental investment because they serve to ensure a suitable marriage partner for a child who will also serve as an important source of investment in grandchildren (Shenk 2007).

The two primary forms of marriage payments are bridewealth, which is paid by the groom or his family to the parents of the bride, and dowry, which is paid by the family of the bride to the new couple or the parents of the groom (Goody & Tambiah 1973). From the perspective of parental investment theory, bridewealth makes daughters cheaper to raise because parents expect to gain from their marriages, whereas dowry makes them more expensive. In bridewealth systems, sons compete with each other, and in polygynous societies with their fathers, for access to the animals or other resources needed to pay bridewealth; in dowry systems, daughters compete with each other and with other family expenditures for the wealth needed to pay dowry.

Among the Kipsigis agro-pastoralists of Tanzania, Borgerhoff Mulder (1998) finds a pattern of same-sex competition and opposite-sex cooperation with regard to marriage payments. Parents use the bridewealth gained from daughters to help fund the marriages of their brothers, so there is competition among sons for bridewealth payments. Later-born sons with many brothers marry at older ages, wed less preferred spouses, make lower bridewealth payments, and have fewer surviving offspring than sons with fewer older brothers, independently of family wealth. In contrast, sons with many sisters do better on all of these measures. Women do not show such tradeoffs with their sisters. In modern India, in contrast, the necessity of paying large dowries for daughters is often cited as a reason for neglect of higher-birth order daughters, whereas sons are more highly valued because of their ability to earn income, care for parents, and gain dowry (Shenk 2007). Both the Kipsigis and Indian cases provide examples of local resource competition and local resource enhancement operating simultaneously.

2.6 Conclusions

This chapter has reviewed the basics of parental investment theory, including some of the key questions to which it has been applied in humans, but it has barely scratched the surface of the literature on this topic. The study of parental investment has been one of the most active areas of enquiry among evolutionary researchers during the last twenty years (Smith & Winterhalder 2000). The reasons are quite simple: reproduction is the most fundamental of evolutionary behaviors, yet human parents face especially complex tradeoffs when deciding how many children to have and how much to invest in each. The wide variety of ecologies and economic systems in which humans live, coupled with the complexity of human social systems, results in a multitude of behavioral options for individuals. Straightforward rules of optimization, such as investing more in the sex of offspring that is likely to have more children, lead to a great deal of variation in observed behaviors across different types of societies and the different social strata within them. When coupled with other sources of variation, such as different marriage systems and the presence or absence of heritable wealth, the potential complexity of human parental investment behavior can seem staggering.

Yet finding simple rules to explain complex behaviors is a hallmark of the evolutionary ecology approach (Smith & Winterhalder 1992), and evolutionary researchers have made great progress in understanding many of the topics discussed in this chapter. Nonetheless plenty of unanswered — or partially answered — questions remain. For instance, many authors have recently emphasized the importance of conflict among kin to understanding parental investment (see, e.g., Borgerhoff Mulder 2007; Lawson & Mace 2009; Leonetti et al. 2007). There has also been a recent resurgence of work on the demographic transition emphasizing cultural influences, risk, and comparative research (see, e.g., Richerson & Boyd 2005; Shenk 2009; Winterhalder & Leslie 2002). New areas of enquiry, such as the study of arranged marriage as a form of parental investment and a site for parent–offspring conflict, are emerging (see, e.g., Apostolou 2007; Buunk et al. 2008). In the last few decades, evolutionary research on human behavior has become more methodologically rigorous, moving from descriptive work to complex simulations and statistical analyses. The number of researchers doing evolutionary work, and the number of cultures they study, has also increased rapidly. As has been true in the past, however, the study of parental investment is likely to remain one of the mainstays of evolutionary approaches to human behavior for many years to come.

References

Alexander R (1990) How did humans evolve? Reflections on the uniquely unique species. Museum of Zoology, The University of Michigan, Special Publication No. 1, Ann Arbor

Alexander RD & Borgia G (1979) On the origin and basis of the male–female phenomenon. In Blum M F, Blum N (eds) *Sexual Selection and Reproductive Competition in Insects*. Academic Press, New York

Alvarez HP (2000) Grandmother hypothesis and primate life histories. American Journal of Physical Anthropology 113:435–450

Al-Jasmi F, Al-Mansoor F, Alsheiba A, Carter AO, Carter TP, Hossain M (2002) Effect of interpreganancy interval on risk of spontaneous preterm birth in Emirati women, United Arab Emirates. Bulletin of World Health Organization 80:871–875

Apostolou M (2007) Sexual selection under parental choice: The role of parents in the evolution of human mating. Evolution and Human Behavior 28:403–409

Ball HL, Hill CM (1996) Reevaluating 'twin infanticide'. Current Anthropology 37(5):856–863

Bateman AJ (1948) Intra-sexual selection in *Drosophila*. Heredity 2:349–368

Bereczkei T, Dunbar RIM (1997) Female-biased reproductive strategies in a Hungarian Gypsy population. Proceedings of the Royal Society of London Series B 264:17–22

Bereczkei T, Dunbar RIM (2002) Helping-at-the-nest and reproduction in a Hungarian Gypsy population. Current Anthropology 43:804–809

Blurton Jones N (1986) Bushman birth spacing: A test for optimal birth intervals. Ethology and Sociobiology 7:91–105

Blurton Jones N (1987) Bushman birth spacing: Direct tests of some simple predictions. Ethology and Sociobiology 8(3):183–203

Borgerhoff Mulder M (1992) Reproductive decisions. In Smith EA, Winterhalder B (eds) *Evolutionary Ecology and Human Behavior*. Aldine de Gruyter, New York

Borgerhoff Mulder M (1998) Brothers and sisters: How sibling interactions affect optimal parental allocations. Human Nature 9(2):119–162

Borgerhoff Mulder M (2007) Hamilton's rule and kin competition: The Kipsigis case. Evolution and Human Behavior 28:299–312

Boone JL (1986) Parental investment and elite family structure in preindustrial states: A case study of late medieval–early modern Portuguese genealogies. American Anthropologist 88(4):859–878

Brown GR, Laland KN, Borgerhoff Mulder M (2009) Bateman's principles and human sex roles. Trends in Ecology and Evolution 24(6):297–304

Buunk AP, Park JH, Dubbs SL (2008) Parent–offspring conflict in mate preferences. Review of General Psychology 12(1):47–62

Cameron EZ, Dalerum F (2009) A Trivers–Willard effect in contemporary humans: Male-biased sex ratios among billionaires. PLoS ONE 4(1):e4195.

Cant MA, Johnstone RA (2008) Reproductive conflict and the separation of reproductive generations in humans. PNAS 105(14):5332–5336.

Chisholm JS (1993) Death, hope, and sex: Life-history theory and the development of reproductive strategies. Current Anthropology 34:1–24

Clark AB (1978) Sex ratio and local resource competition in a prosimian primate. Science 201(4351):163–165

Clarke AL, Low BS (2001) Testing evolutionary hypotheses with demographic data. Population and Development Review 27:633–660

Clutton-Brock TH (1989): Mammalian mating systems. Proceedings of the Royal Society of London B 236:339–372

Clutton-Brock TH (1991) *The Evolution of Parental Care*. Princeton University Press, Princeton, NJ

Clutton-Brock TH, Godfray C (1991) Parental investment. In Krebs JR, Davies NB (eds) *Behavioural Ecology: An Evolutionary Approach*. Blackwell, Boston

Cronk L (1991) Preferential parental investment in daughters over sons. Human Nature 2:387–417

Daly M, Wilson M (1984) A sociobiological analysis of human infanticide. In Hausfater G, Hrdy SB (eds) *Infanticide: Comparative and Evolutionary Perspectives*. Aldine, New York

Daly M, Wilson M (1988) *Homicide*. Aldine De Gruyter, Hawthorne, NY

Deady DK, Law Smith MJ, Kent JP, Dunbar RIM (2007) Is priesthood an adaptive strategy? Evidence from a historical Irish population. Human Nature 17(4)393–404

Dickemann M (1979) The ecology of mating systems in hypergynous dowry societies. Social Science Information 18:163–196

Draper P, Harpending H (1982) Father absence and reproductive strategy: An evolutionary perspective. Journal of Anthropological Research 38:255-273

Emlen ST, Emlen JM, Levin SA (1986) Sex-ratio selection in species with helpers-at-the-nest. The American Naturalist 127(1):1–8

Finer LB, Frohwirth LF, Dauphinee LA, Singh S, Moore AM (2005) Reasons US women have abortions: Quantitative and qualitative perspectives. Perspectives on Sexual and Reproductive Health 37(3):110–118

Fisher RA (1930) *The Genetical Theory of Natural Selection*. Oxford University Press, Oxford

Fisher RA (1958) *The Genetical Theory of Natural Selection*, 2nd edn. Dover, New York

Flinn M, Quinlan RL, Ward CV, Coe MK (2007) Evolution of the human family: Cooperative males, long social childhoods, smart mothers, and extended kin networks. In Salmon C, Shackelford T (eds) *Family Relationships*. Oxford University Press, Oxford

Fox M, Sear R, Beise J, Ragsdale G, Voland E, Knapp LA (2009) Grandma plays favourites: X-chromosome relatedness and sex-specific childhood mortality. Proceedings of the Royal Society B 277(1681):567–573

Fuster V (1984) Extramarital reproduction and infant mortality in rural Galicia (Spain). Journal of Human Evolution 13(5):457–463

Gabler S, Voland E (1994): Fitness of twinning. Human Biology 66:699–713

Gasser M, Kaiser M, Berrigan D, Stearns SC (2000) Life history correlates of evolution under high and low adult mortality. Evolution 54:1260–1272

Gaulin SJC, McBurney DH, Brakeman-Wartell SL (1997) Matrilateral biases in the investment of aunts and uncles: A consequence and measure of paternity uncertainty. Human Nature 8(2):139-151

Geary DC (2000) Evolution and proximate expression of human paternal investment. Psychological Bulletin 126:55–77

Gibson MA, Mace R (2005) Helpful grandmothers in rural Ethiopia: A study of the effect of kin on child survival and growth. Evolution and Human Behavior 26:469–482

Goody J, Tambiah SJ (1973) *Bridewealth and Dowry. Cambridge Papers in Social Anthropology No. 7*. Cambridge University Press, Cambridge

Gowaty PA, Lennartz MR (1985) Sex ratios of nestling and fledgling red-cockaded woodpeckers (*Picoides borealis*) favor males. The American Naturalist 126(3):347–353

Hagen EH, Barrett HC, Price ME (2006) Do human parents face a quantity–quality tradeoff? Evidence from a Shuar community. American Journal of Physical Anthropology 130:405–418

Hajnal J (1965) European marriage patterns in perspective. In Glass DV, Eversley DEC (eds) *Population in History, Essays in Historical Demography*. Aldine: Chicago

Hamilton WD (1963) The evolution of altruistic behavior. American Naturalist 97:354–356

Hamilton WD (1964) The genetical evolution of social behavior. Journal of Theoretical Biology 7:1–52

Hamilton WD (1966) The moulding of senescence by natural selection. Journal of Theoretical Biology 12(1):12–45

Hamilton WD (1967) Extraordinary sex ratios. Science 156:477–488

Harrell S (1997) *Human Families*. Westview Press, Boulder, Colorado

Harrison KA, Fleming AF, Briggs ND, Rossiter CE (1985) Growth during pregnancy in Nigerian teenage primigravidae. British Journal of Obstetrics and Gynaecology 92 Supplement 5:32–39

Hawkes K, O'Connell JF, Blurton-Jones NG (1997) Hadza women's time allocation, offspring provisioning, and the evolution of long post-menopausal lifespans. Current Anthropology 38:551–578

Hewlett B (1991) *Intimate Fathers: The Nature and Context of Aka Pygmy Paternal Infant Care*. University of Michigan Press, Ann Arbor

Hill EM, Low BS (1991) Contemporary abortion patterns: A life history approach. Ethology and Sociobiology 13:35–48

Hill K (1993) Life history theory and evolutionary anthropology. Evolutionary Anthropology 2:78–88

Hill K, Hurtado M (1991) The evolution of premature reproductive senescence in human females: An evaluation of the 'grandmother hypothesis'. Human Nature 2:313–350

Hill K, Hurtado M (1996) *Ache Life History: The Ecology and Demography of a Foraging People*. Aldine de Gruyter, New York

Hill K, Kaplan H (1999) Life history traits in humans: Theory and empirical studies. Annual Review of Anthropology 28:397–430

Hobcraft JN, McDonald JW, Rutstein SO (1985) Demographic determinants of infant and early child mortality: A comparative analysis. Population Studies 39(3):363–385

Hopcroft R (2005) Parental status and differential investment in sons and daughters: Trivers–Willard revisited. Social Forces 83(3):1111–1136

Hrdy S (1992) Fitness tradeoffs in the history and evolution of delegated mothering with special reference to wet-nursing, abandonment, and infanticide. Ethology and Sociobiology 13:409–442

Hrdy S (1999) *Mother Nature: A History of Mothers, Infants, and Natural Selection*. Pantheon, New York, NY

Hrdy S (2009) *Mothers and Others: The Evolutionary Origins of Mutual Understanding*. Belknap Press of Harvard University Press, Cambridge, MA

Kaplan H (1996) A theory of fertility and parental investment in traditional and modern human societies. Yearbook of Physical Anthropology 39:91–135

Kaplan H, Hill K, Lancaster JL, Hurtado AM (2000) A theory of human life history evolution: Diet, intelligence, and longevity. Evolutionary Anthropology 9:156–185

Kaplan H, Lancaster JB, Tucker WT, Anderson KG (2002) Evolutionary approach to below re-placement fertility. American Journal of Human Biology 14:233–256

Kramer KL (2005) Children's help and the pace of reproduction: Cooperative breeding in humans. Evolutionary Anthropology 14:224–237

Kramer KL, Boone JL (2002) Why intensive agriculturalists have higher fertility: A household energy budget approach. Current Anthropology 43(3):511–517

Lawson D, Mace R (2009) Tradeoffs in modern parenting: A longitudinal study of sibling com-petition for parental care. Population Index 30(3):170–183

Lee R (1979) *The !Kung San: Men, Women, and Work in a Foraging Society*. Cambridge Univer-sity Press, Cambridge, UK

Lee R and Devore I (1968) *Man the Hunter: The First Intensive Survey of a Single, Crucial Stage of Human Development — Man's Once Universal Hunting Way of Life*. Aldine De Gruyter, Hawthorne, NY

Leonetti DL, Nath DC, Hemam NS (2007) 'In-law conflict': Women's reproductive lives and the roles of their mothers and husbands among the matrilineal Khasi (with comments). Current Anthropology 48:861–890

Lessels CM (1991) The evolution of life histories. In Krebs JR, Davies NB (eds) *Behavioural Ecology: An Evolutionary Approach*, 3rd edn. Blackwell Scientific Publications, Oxford

Low BS (1991) Reproductive life in 19th century Sweden: An evolutionary perspective on demo-graphic phenomena. Ethology and Sociobiology 12:411–468

Low BS (1998) The evolution of human life histories. In Crawford C, Krebs DL (eds) *Handbook of Evolutionary Psychology: Ideas, Issues, and Applications*. Lawrence Erlbaum Associates, Mahwah, NJ

Lycett J, Dunbar RIM (1999) Abortion rates reflect the optimization of parental investment strategies. Proceedings of the Royal Society of London Series B 266:2355–2358

MacArthur RH, Wilson EO (1967) *The Theory of Island Biogeography*. Princeton University Press, Princeton, NJ

Mace R (1996) Biased parental investment and reproductive success in Gabbra pastoralists. Behavioral Ecology and Sociobiology 38:75–81

Mace R (2000) Evolutionary ecology of human life history. Animal Behaviour 29:1–10

Mace R, Sear R (2005) Are humans co-operative breeders? In Voland E, Chasiotis A, SchiefenHoevel W (eds) *Grandmotherhood: The Evolutionary Significance of the Second Half of Female Life*. Rutgers University Press, Piscataway

Marlowe F (2000) Paternal investment and the human mating system. Behavioural Processes 51:45–61

Miller G (2000) *The Mating Mind: How Sexual Choice Shaped the Evolution of Human Nature*. Anchor Books, New York

Nettle D, Pollet TV (2008) Natural selection on male wealth in humans. American Naturalist 172:658–666

Quinlan R (2007) Human parental effort and environmental risk. Proceedings of the Royal Society B: Biological Sciences 274:121–125

Quinlan R, Quinlan M (2007) Parenting and cultures of risk: A comparative analysis of infidelity, aggression, and witchcraft. American Anthropologist 109:164–179

Pashos A, McBurney D (2008) Kin relationships and the caregiving biases of grandparents, aunts, and uncles. Human Nature 19(3):311–330

Rogers AR (1993) Why menopause? Evolutionary Ecology 7:406–420

Scelza BA (2010) Father's presence speeds the social and reproductive careers of sons. Current Anthropology 51(2):295–303

Scheper-Hughes N (1992) *Death Without Weeping: The Violence of Everyday Life in Brazil*. University of California Press, Berkeley, CA

Sear R, Shanley D, McGregor IA, Mace R (2001) The fitness of twin mothers: Evidence from rural Gambia. Journal of Evolutionary Biology 14:433–443

Shenk MK (2005) Kin networks in wage-labor economies: Effects on child and marriage market outcomes. Human Nature 16:81–114

Shenk MK (2007) Dowry and public policy in contemporary India: The behavioral ecology of a social 'evil'. Human Nature 18(2):242–263

Shenk MK (2009) Testing three evolutionary models of the demographic transition: Patterns of fertility and age at marriage in Urban South India. American Journal of Human Biology 21:501–511

Sieff DF (1990) Explaining biased sex ratios in human populations: A critique of recent studies. Current Anthropology 31(1):25–48

Silk JB (1980) Adoption and kinship in Oceania. American Anthropologist 82:799–820

Silk JB (1983) Local resource competition and facultative adjustment of sex ratios in relation to competitive abilities. The American Naturalist 121(1):56–66

Smith EA, Winterhalder B (1992) Evolutionary Ecology and Human Behavior. Aldine de Gruyter, New York

Smith EA, Winterhalder B (2000) Analyzing adaptive strategies: Human behavioral ecology at twenty-five. Evolutionary Anthropology 9:51–72

Strassmann BI, Gillespie B (2002) Life-history theory, fertility, and reproductive success in humans. Proceedings of the Royal Society of London Series B 269:553–562

Trivers RL (1972) Parental investment and sexual selection. In Campbell B (ed) *Sexual Selection and the Descent of Man 1871–1971*. Aldine, Chicago, IL

Trivers RL (1974) Parent–offspring conflict. American Zoologist 14:249–64

Trivers RL (1985) *Social Evolution*. Benjamin/Cummings Publishing Group, Menlo Park California

Trivers RL, Willard DE (1973) Natural selection of parental ability to vary the sex ratio of offspring. Science 179:90–92

Tullberg BS, Lummaa V (2001) Induced abortion ratio in modern Sweden falls with age, but rises again before menopause. Evolution and Human Behavior 22(1):1–10

Turke P (1988): Helpers at the nest: Childcare networks on Ifaluk. In Betzig LL, Borgerhoff Mulder M, Turke P (eds) *Human Reproductive Behavior: A Darwinian Perspective*. Cambridge University Press, Cambridge

Van Schaik C, Hrdy SB (1991) Intensity of local resource competition shapes the relationship between maternal rank and sex ratios at birth in Cercopithecine primates. The American Naturalist 138(6):1555–1562

Voland E, Chasiotis A, Schiefenhovel W (2005) *Grandmotherhood: The Evolutionary Significance of the Second Half of Female Life*. Rutgers University Press, Rutgers, NJ

Voland E, Dunbar RIM (1995) Resource competition and reproduction — the relationship between economic and parental strategies in the Krummhörn population (1720–1874). Human Nature 6:33–49

Voland E, Dunbar RIM, Engel C, Stephan P (1997) Population increase and sex-biased parental investment: Evidence from 18th and 19th century Germany. Current Anthropology 38:129–135

Voland E (1998) Evolutionary ecology of human reproduction. Annual review of anthropology 27:347–374

Williams GC (1957) Pleiotropy, natural selection, and the evolution of senescence. Evolution 11:398–411

Williams GC (1966) *Adaptation and Natural Selection: A Critique of some Current Evolutionary Thought*. Princeton University Press, Princeton, NJ

Wilson M, Daly M (1997) Life expectancy, economic inequality, homicide, and reproductive timing in Chicago neighborhoods. British Medical Journal 314:1271–1274

Winterhalder B, Leslie P (2002) Risk-sensitive fertility: The variance compensation hypothesis. Evolution and Human Behavior 23:59–82

Wood JW, O'Connor KA, Holman DJ, Brindle E, Barsom SH, Grimes MA (2001) The evolution of menopause by antagonistic pleiotropy. Working Paper 01–04, Center for Studies in Demography & Ecology, University of Washington

Chapter 3
Our Social Roots: How Local Ecology Shapes Our Social Structures

Ruth Mace

Abstract There is overwhelming evidence that wide-ranging aspects of human biology and human behavior can be considered as adaptations to different subsistence systems. Wider environmental and ecological correlates of behavioral and cultural traits are generally best understood as being mediated by differences in subsistence strategies. Modes of subsistence profoundly influence both human biology, as documented in genetic changes, and human social behavior and cultural norms, such as kinship, marriage, descent, wealth inheritance, and political systems. Thus both cultural and biological factors usually need to be considered together in studies of human evolutionary ecology, combined in specifically defined evolutionary models. Models of cultural adaptation to environmental conditions can be subjected to the same or similar tests that behavioral ecologists have used to seek evidence for adaptive behavior in other species. Phylogenetic comparative methods are proving useful, both for studying co-evolutionary hypotheses (cultural and/or gene–culture co-evolution), and for estimating ancestral states of prehistoric societies. This form of formal cross-cultural comparison is helping to put history back into anthropology, and helping us to understand cultural evolutionary processes at a number of levels.

3.1 Adaptation and Maladaptation

Humans are an extremely successful species, able to inhabit almost any environment on earth. The way in which we do that undoubtedly has to do with a range of sophisticated subsistence strategies. We probably developed these diverse strategies with the help of our superior cognitive powers, especially our ability to observe the behavior of others, and learn from it, and then possibly improve on it, and pass that knowledge on to our descendents. Cumulative cultural evolution saves us the costs of figuring everything out for ourselves each generation and allows us to benefit

Ruth Mace
Department of Anthropology, University College London, Gower Street, London WC1E 6BT, United Kingdom, e-mail: r.mace@ucl.ac.uk

U.J. Frey et al. (eds.), *Essential Building Blocks of Human Nature*, The Frontiers Collection, DOI 10.1007/978-3-642-13968-0_3, © Springer-Verlag Berlin Heidelberg 2011

from the discoveries of earlier generations, and this undoubtedly gave us the edge over other, non-cultural species.

Given that it is blindingly obvious that culture has been so useful, it is perhaps surprising that the majority of cultural anthropologists do not consider most cultural behavior to be adaptive — certainly not in the Darwinian sense, at any rate (Laland & Brown 2002; Perry & Mace 2010; Segerstrale 2000). And whilst most evolutionary anthropologists would consider the ability to be a cultural species and learn from others as an adaptation, probably as much attention has been focused on explaining how cultural transmission (or social learning) could generate apparently maladaptive aspects of human behavior rather than focusing on its benefits in helping humans adapt to environments. Relatively few studies have tried to test adaptive hypotheses about behavioral or cultural diversity at the macro-level, which is the topic of this chapter.

Adaptations are designed by natural selection to maximize inclusive fitness. Behavioral ecologists use three main approaches to test adaptive hypotheses about the evolution of behavior: these are experimentation, testing the predictions of optimality models, and the comparative method. When a particular adaptive model fails to explain observed phenomena, the usual modus operandi is to seek a better model, assuming that some vital cost or benefit has been wrongly incorporated or overlooked; hence our understanding of the evolutionary basis of that behavior is enhanced by ruling out multiple alternative explanations.

There are a number of reasons why behavior may not be adaptive. The most important one is that a rapid change in the environment will cause temporary maladaptation, as evolution takes time to work. This is often referred to as a 'mismatch' argument, or an example of 'evolutionary lag'. If the proximate mechanisms for social learning or other determinants of behavior (such as preferences) evolved in environmental conditions that are no longer current, then emergent behavior may no longer promote fitness. Unfortunately these mismatch arguments are difficult to test; quite apart from proving a negative, it is difficult to establish what the costs and benefits of behavior were in an environmental or cultural context that no longer exists.

How long it takes for behavioral adaptation to evolve is not clear. A number of recent studies have given us a window on the pace of genetic evolution in the face of cultural changes in subsistence practices. Lactose tolerance evolved amongst those keeping livestock for dairy (Bersaglieri et al. 2004; Ingram et al. 2009), alleles protective against prion-based neurodegenerative disease (kuru) in the Fore of New Guinea have been selected for by cannibalism (Mead et al. 2009), and the frequency of alleles associated with alcohol dehydrogenase appear to map the history of rice cultivation in south Asia (Peng et al. 2010). These all provide demonstrations of recent strong selection causing rapid evolution, occurring within the last few thousand years or less, in genetic traits associated with changes in subsistence strategy and diet. Of course, we are unlikely ever to find such clear signatures of specific genes influencing behavior, behavioral genetics being altogether more complex (Plomin 2001). The heritability of behavioral traits is very hard to measure. Nonetheless it would seem unlikely that behavioral adaptation is slower than digestive adaptation. Evolutionary psychologists who assert that our behavior is adapted to Pleistocene conditions have faced opposition (Laland et al. 2002). Cultural evolution can be much faster than genetic evolution, so mismatch arguments for maladaptation in cultural traits are perhaps on even shakier ground than mismatch arguments about genetic maladaptation. Some have argued that cultural evolution may in fact have caused genetic evolution to accelerate (Hawks et al., 2007), perhaps by generating so many new niches. Alternatively, niche construction by humans could be a mechanism by which we can avoid mismatches between their environment and their optimal living conditions (Laland & Brown 2006; Odling-Smee, Laland, & Feldman 2003).

Some evolutionary anthropologists take the position that cultural inheritance mechanisms can generate stable outcomes that result in behaviors that are not necessarily adaptive in the genetic sense, because generalized social learning rules may be used which, in some circumstances, happen to promote the spread of that cultural trait but not necessarily the inclusive fitness of the person performing that behavior. Social learning enables cultural traits to move between individuals in a non-Mendelian way. Many cultural traits are copied directly from biological parents, but it is also true that learning might involve a range of possible 'cultural parents' chosen on grounds of frequency of contact, proximity, prestige, efficacy, or any other criteria, often referred to as biased transmission (Boyd & Richerson 1985). Variation in the possible modes of cultural transmission can therefore influence the types and dynamics of cultural behaviors that evolve. For example, the transmission mode of conformist bias (copying the common cultural traits in your group) can cause cultural groups to resist invasion by mutant cultural types. This could allow between-group variation to be maintained long enough to be subject to cultural group selection; this might favor the evolution of traits that favor the group (Richerson & Boyd 2005). The cultural evolution of traits that spread through benefits to the whole group might be rather slow (Soltis, Boyd, & Richerson 1995).

One example of a model of how transmission mechanisms alone are invoked to explain maladaptive behavior is due to Tanaka et al. (2009), who explore the role of social learning mechanisms in explaining the persistence of (self-prescribed) medical treatments that have no efficacy at all: so-called 'traditional', 'alternative', and even some modern medical treatments. In this example, individuals are assumed to copy self-medication treatments in proportion to the rate at which they observe those treatments being used by other individuals suffering similar medical conditions to themselves. This very interesting paper makes some counter-intuitive predictions, including this, for example: if a user takes the treatment for longer because the illness does not get better, then the opportunity for her to become a model for other social learners increases (Tanaka, Kendal, & Laland 2009). This means that behavior might persist because social learning is generally more effective than trial and error, but can lead to the copying of harmful traits in some circumstances. This is essentially a proximate explanation for why a harmful or neutral behavior might persist over time.

However, one might expect humans to improve their learning mechanism, perhaps by applying a more sophisticated rule about when to use social learning and when not to, which could enhance their inclusive fitness in the long run. Thus the explanation for the persistence of the use of ineffective medical treatments becomes based either on constraints (the task at hand is simply beyond the capacity of the human mind to resolve) or a classic mismatch or evolutionary lag argument; thus it is a more sophisticated version, but philosophically not that dissimilar from the verbal arguments commonly employed by evolutionary psychologists to explain much of our apparently maladaptive behavior. This is not to say the model does not provide a convincing proximate explanation for the observed phenomena of useless self-medication. Such cultural evolutionary models have as yet rarely been parameterized by fitting to datasets from real behavior, but only supported by the observation that the general phenomena described does exist. So as yet it is hard to know how frequent such cases of truly maladaptive behavior, arising due to social learning, really are.

3.2 Testing Hypotheses About Adaptation in Human Cultural Behavior Through Cross-Cultural Comparison

There is no theoretical reason why the study of human cultural adaptation should not be investigated in roughly the same manner as behavioral ecologists seek adaptation in the natural world; although human studies can present additional challenges. In anthropology, experimental manipulation of human subsistence is rarely possible in a naturalistic setting. Sometimes it is possible to make use of development interventions or other such changes to find 'natural experiments' (e.g., Gibson & Mace 2006). Optimality models are very useful, and have been used to show how human behavior can be understood as adaptive in certain environments in a number of domains, especially to foraging theory and reproductive behavior (topics beyond the scope of this paper, which are discussed elsewhere in this volume). These approaches use individual-level variation within populations. Such individual-level effects can also explain wider cultural differences, although cultural differences are, almost by definition, a property of the group (culture) rather than a property of the individual; and individual deviation from some cultural norms can be strongly suppressed (by legal restriction, ostracism, or other methods). So individual-level variation in cultural behavior is not necessarily what one is trying to explain when interpreting cultural differences. Then a cross-cultural comparative method becomes a key tool. Cross-cultural comparison was indeed the historical basis of anthropology.

3.2.1 Ecological Correlates of Human Social Behavior

An intuitively appealing method with which to understand ecological adaptation is to examine which human social traits co-vary with ecological variables across cultures. In a recent review of a number of such studies, most were correlates of parasite prevalence and/or latitude (Nettle 2009). Of course, latitude itself correlates with parasite load, as there are more species near the tropics, including parasites. However latitudinal gradient in cross-cultural human ecology is a bit like socioeconomic status within human populations: nearly everything correlates with it, and it is very hard to control for fully. Many of the studies listed in Nettle (2009) are in danger of serious misinterpretation for this and other reasons. For example, polygynous marriage, promiscuous socio-sexuality, high fertility, and a more female-biased sex ratio are all more common in the tropics where there are more parasites. Is this due to parasites, due to different subsistence strategies related to other aspects of ecological difference or to cultural history being different in Africa and Europe, or due to economic development that for various reasons has occurred more in the north than in the south, or for other reasons (Diamond 1997)? For example, Mace and Jordan found that female-biased sex ratio at birth correlated with high fertility and mortality rates, even after controlling for phylogenetic relationships between groups (Mace & Jordan 2005) and interpreted it in terms of high costs of reproduction causing fewer males, in line with evolutionary life history theory. Of course, high fertility and mortality co-vary strongly with economic development (and progress of the demographic transition) and parasite load and latitude, so an association with high fertility could underlie the geographic patterns that generate correlations with all these variables. Over-interpretation of ecological correlation is nothing new; confusing correlation with causation in earlier attempts at formal cross-cultural comparison based on the *Human Relations Area Files* from the 1960s on may have contributed to cultural anthropologists becoming so sceptical about quantitative methods that they all but abandoned them.

The difficulty in controlling for cultural and biological history is discussed below (see Sect. 3.2.3). But spurious correlation due to the non-independence of cultural data points is only one dimension of the wider problem with these correlational studies, which is that they are not explicit enough about the evolutionary models that generate the associations observed. One of the advantages of some phylogenetic comparative methods is that they enable us to discern between explicitly defined alternative evolutionary models. Second, and related to the above, most studies of ecological correlates of behavior do not address differences in subsistence system. Subsistence systems influence how human populations get resources from their environment, and we already know that they have a profound influence on human social systems and hence behavior.

One recent study that both considers changes in subsistence and formally evaluates explicit evolutionary models, and is also an exception to the general rule that cultural or gene–cultural co-evolutionary models are not fitted to real data, is the simulation by Itan et al. (2009) of the spread of agriculture and lactase persistence across Europe. It presents a gene–culture co-evolutionary model of the emergence of lactose tolerance (or lactase persistence into adulthood) as an adaptation to milk-drinking, in a population where individuals can switch between gathering, farming, and pastoralism. Lactase persistence shows a strong latitudinal gradient in Europe, which on the face of it supports the hypothesis that it is selected for in ecological conditions with low levels of sunshine due to vitamin D deficiency (Flatz & Rot-thauwe 1973). Itan et al. fit some of their model parameters explicitly, by using Bayesian inference (Beaumont et al. 2002) to determine which parameters of the model best predict the present day distribution of the allele associated with lactose tolerance in Europeans (known as -13910-T). This exercise in statistical inference not only locates the likely starting point (time and place) of this gene–cultural co-evolutionary process in central Europe, but also shows that the latitudinal gradient in the T allele is not due to stronger selection at high latitudes but simply due to the demographic history of the wave of expansion generated by an increasing density of farmers taking over new territory to the north (Itan et al. 2009). The genes for lactase persistence ride on the crest of the wave of advance of territories occupied by the new subsistence strategy, rather than work their way back into existing populations. Holden and Mace also found no evidence for the vitamin D hypothesis for lactase persistence using a global cross-cultural sample and a phylogenetic comparative method (Holden & Mace 1997). Itan et al. show that a model based on demic expansion best explains the patterns of the allele distribution observed today (which, incidentally, they estimate has not yet reached equilibrium). Hence both proximate models of emergence and ultimate adaptive function are addressed together in a co-evolutionary model of subsistence change and human biology.

3.2.2 How Social Behavior is Adapted to Subsistence Strategies

Changes in subsistence strategy were responsible for most of the major evolutionary transitions in human societies, particularly the advent of agriculture coinciding with changes in marriage, descent, fertility, and other social traits which also promoted greatly increased population densities and increase in social inequality. The behavioral ecology of all these co-adaptive changes is now reasonably well understood.

Hunter–gatherers live in bands, with largely monogamous marriage, relatively low fertility, no heritable wealth to speak of, and relatively egalitarian social systems. Since the adoption of agriculture, human social systems have been largely shaped by the existence of exceedingly important resources (such as fields or livestock) that can be controlled or owned (by individuals or by groups), and passed down to future generations; access to such resources greatly influences the future reproductive success of descendents and generates inequalities in wealth and political power (Kaplan, Hooper, & Gurven 2009). Population densities increased with the advent of agriculture and more complex political systems emerged, correlating with human ethno-linguistic groups becoming larger and more politically complex (Currie & Mace 2009; Johnson & Earle 2000). Systems of wealth inheritance are fundamentally linked with systems of marriage and the associated transfers of wealth at marriage, and thus marriage and descent systems are products of the socioeconomic system on which societies are based. As is well known to behavioral ecologists, if males are able to monopolize access to a bit of land that has the resources required for breeding, then that resource can be used to attract females, who will mate polygynously, if need be, to acquire that resource. Thus resource-defence polygyny, not dissimilar to that described in birds (Orians 1969), is also common in humans. As in other species, such polygynous systems can only really emerge where there are sufficient resources for females to raise their children without a great deal of individual help from fathers. Resources such as livestock are particularly associated with polygynous marriage, and male-biased wealth inheritance (Hartung 1982). If the number of grandchildren can be enhanced more by leaving livestock to sons (enabling them to marry earlier and more often) than to daughters, which is the case under resource-based polygyny, then patrilineal wealth inheritance norms doing just that will emerge (Mace 1996).

Within lineal family systems, patriliny is by far the most common, but a significant minority (about 17%) of systems described in the *Ethnographic Atlas of World Cultures* (Murdock, 1967) are matrilineal. Marriage bonds are often weak in matrilineal systems, with women frequently marrying several husbands over the course of their lives, as resources are passed down the female line. The ecology that is predictive of matriliny is based on systems where resources cannot be easily monopolized by males to attract females. In Africa, it is strongly associated with the absence of livestock (Aberle 1961; Holden & Mace 2003). African crop production is often not land limited but labor limited, so, whereas livestock offers females the promise of resources relatively easily accumulated, land of the type that is only of value after months of back-breaking labor in the fields does not generally provide men with the opportunity to monopolize large areas to attract mates. Women will only remain married to men as long as they help them work the land. In other parts of the world matriliny has been proposed to be associated with high male mortality rates (either in warfare, as in some matrilineal Native American groups, or with ocean fishing as in the Pacific). Whatever the underlying ecology, women in matrilineal systems rely on mothers, daughters, and sisters to support their family, as help from males is often transitory. Paternity uncertainty tends to be high in matrilineal systems, although the extent to which this is a cause or consequence of matrilineal descent systems is a matter of debate (Hartung 1985). In the case of correlations between subsistence, descent, and kinship, understanding of how fitness is maximized at the individual level helps explain larger scale cross-cultural patterns.

3.2.3 Cultural Phylogenetics

Elsewhere we have argued that phylogenetic comparative methods are an appropriate formal comparative method to use in anthropology (Mace & Pagel 1994), just as they are in evolutionary biology in general (Harvey & Pagel 1991). Phylogenetic comparative methods take into account the fact that cultures are not independent of each other, and, in a manner analogous to biological evolution, daughter cultures evolve from mother cultures, generating a tree-like pattern of origin, or a phylogeny. Whilst a perfect phylogeny may not be a perfect model for the evolution of cultures (as indeed it sometimes is not even for the evolution of some species), it is generally a far better model than the model on which other general statistical methods rely, i.e., assuming that all societies are related to each completely equidistantly. Ignoring the ancestor–descendent relationships between cultures can generate significant errors (Harvey & Pagel 1991). Furthermore, a powerful set of statistical tools have been developed by evolutionary biologists to examine various evolutionary processes on phylogenies, which go beyond just seeking correlation to examine underlying evolutionary processes (Pagel 1999). In recent years, we and others have been applying this toolkit to examine cultural evolution.

The need to build cultural phylogenetic trees to use phylogenetic comparative methods was also partly responsible for a resurgence of interest in inferring historical patterns of human migrations. Trees tracking human population history have been built using comparison of elements of language. Comparing elements of core vocabulary to infer trees of population history has been especially productive in some large language families (notably Bantu, Indo-European, and Austronesian), where the trees generated fit well with what linguists, archaeologists, and historians believe to be realistic models of population spread (Mace, Holden, & Shennan 2005). Archaeologists have also applied these techniques to aspects of material culture (Lipo 2006; Mace et al. 2005). Phylogenetic reconstructions using linguistic data have enabled us to arbitrate between different historical migration proposals in cases that genetic, archaeological, and other data or methods have not enabled us to distinguish (Gray & Atkinson 2003; Gray & Jordan 2000). It is probably not a coincidence that all three of these language families have had relatively recent dispersals, largely based on technological advances enabling them to advance successfully into new territories. Language trees may have such a strong phylogenetic signal because language is a neutral trait (i.e., the forms of words themselves have no fitness implications) and has twin pressures on it that maintain distinct but consistent forms. Those from within include conformist bias (or frequency dependence); you and your children have to speak the language most of those around you are speaking if you are to succeed. Those from without are forces that act to maintain group boundaries, to signal difference from and promote mutual unintelligibility with one's neighbors. When migrants enter new groups they may pass their genes into their new population but they do not usually pass on their language. Hence linguistic trees tend to return a much stronger phylogenetic signal than do genetic trees within the human species.

Building phylogenetic trees is thus the 'first step' for evolutionary anthropologists who want to test cultural hypotheses using phylogenetic comparative methods. The first use of phylogenetic comparative methods in anthropology was to examine the co-evolution of cultural traits, or cultural and biological traits. Whereas simple regressions across cultures, not accounting for phylogeny, can generate spurious correlations, phylogenetic comparative methods seek evidence for the fact that change in one character on the tree is associated with change in another character, hence providing evidence that the two traits are functionally linked. The method we have used most often to examine the co-evolution of discrete traits on phylogenies, viz., DISCRETE (Pagel 1994), directly compares different predefined models of evolution, including those in which the evolution of two discrete traits (i.e., taking a value of presence or absence) is correlated, and those in which traits are evolving independently of each other. Maximum likelihood or Bayesian methods (Pagel & Meade 2006) are used to determine which model is most likely to have generated the extant patterns of data observed at the tips of the tree (i.e., in the present) (Pagel 1999). Because models of evolution are specifically defined in these methods, it is possible not simply to look for correlation, but also to estimate the most likely direction of causation. It is possible to ask whether a change in one trait drives the change in another, or vice versa. For example, whilst it has long been known that the people living in cultures with a history of dairying are more likely to be lactose tolerant (Simoons 1972), we were able to use DISCRETE (Pagel 1994) to show that a model in which a shift to keeping cattle preceding a switch to lactose tolerance was a far better fit than a model in which the switch to lactose tolerance occurred before the adoption of dairying — thus providing strong support for the hypothesis that lactose tolerance evolves in direct response to and as an adaptation to milk-drinking (Holden et al. 1997).

Since that early study, we have examined the co-evolution of subsistence systems and social systems such as marriage, kinship, and descent rules. We were able to show that, in Bantu-speaking populations, patrilineal social systems were associated with pastoralism whereas matrilineal systems were associated with a lack of cattle-keeping (Holden et al., 2003). The direction of change confirmed the hypothesis that a transition to pastoralism precedes a switch to patrilineal descent systems (see Box 1 at the end of the section). In other studies it has been shown that monogamous marriage co-evolves with dowry (although in this case the arrow of causation is less clear) in Indo-Europeans (Fortunato, Holden, & Mace 2006; Pagel & Meade 2005).

Other than just examining co-evolution, cultural phylogenetic methods have also been used to infer ancestral states. Phylogenetic techniques rely on using the extant distribution of traits, and the phylogeny, to infer which evolutionary processes were most likely to have generated that distribution (Pagel 1999). This involves attributing a likelihood that any particular node on the tree was at a particular state. In the case of Bayesian methods, the likelihood that that node actually existed (given the uncertainty in the phylogeny) is also taken into account (Pagel & Meade 2006). Hence implicit in the method is the inference of ancestral conditions. In evolutionary biology, this has actually become the purpose for which DISCRETE (Pagel 1994) has been most used, and I suspect a similar trend could emerge in anthropology. Social systems rarely leave any trace in the archaeological record. Although genetic data might throw some light, again using not dissimilar methods of statistical inference, on past human mating patterns, such inferences are usually post-hoc discussion points. But most anthropology and ethnography is really confined to the present and recent history within living memory or, if we are lucky, in written or oral histories. Cultural phylogenetic techniques potentially enable us to put history and even prehistory back into anthropology. We have used these techniques to show that the most likely ancestral conditions of Proto-Malayo-Polynesian ($\sim$ 4500 years ago) was matrilineal, with patrilineal systems evolving later on in the Austronesian family (Jordan et al. 2009). Similarly we have been able to show that dowry and monogamy were probably ancestral in Indo-European (Fortunato et al. 2006). Whilst studies of ancestral condition do not necessarily demonstrate adaptation, they are essential in arbitrating between different causal hypotheses for the origins of cultural traits. For example, if the ancestral Indo-Europeans were monogamous, then monogamy long predates the emergence of Christianity (which is only about 2000 years old), debunking the common assumption that Christianity was the driving force behind monogamy in Europe. It provides support for the notion that prevailing local social systems and conventions generally determine religious rules rather than vice versa.

The Bayesian phylogenetic methods developed by Pagel and Meade (2006) are also general enough to compare evolutionary pathways that may involve several different states, and evaluate the most likely pathways that transitions between these states have followed to generate the patterns we see in the present. Currie et al. (in preparation) use this method to show that political complexity arises through a regular sequence from simple to politically complex societies in Austronesia, although collapse of complexity can follow any sequential or non-sequential route.

Box 1. Using phylogenetic comparative methods to study human cultural evolution: the co-evolution of cattle-keeping and matriliny (Holden and Mace 2003):

(i)   Matrilineal groups in Africa are roughly coincident with an area where cattle are not kept, due to the tetse belt. Holden and Mace (2003) used phylogenetic comparative methods to test whether the social system of patrilineal descent was co-evolving with cattle-keeping in the Bantu-speaking populations. This work illustrates how language data can be used to make phylogenetic trees of population history, at least in some language groups, and that these can be used to test functional hypotheses about co-evolution.

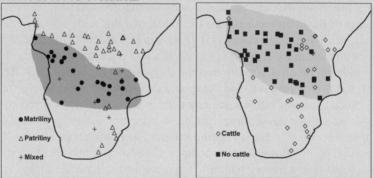

(ii)   This tree was built using language similarity in a maximum parsimony tree-building algorithm to ascertain the historical relationships between groups (Holden and Mace 2003). The phylogenetic groupings match well with those that linguists and archaeologists have identified. The colour of the ethno-linguistic group shows the descent system and whether or not they are part of a cattle-keeping group.

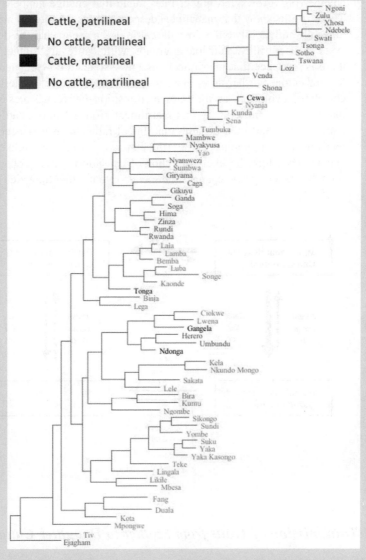

(iii)   Using the phylogenetic comparative method DISCRETE (Pagel 1994), rates of transition between states are estimated, to show that there is co-evolution between descent systems and cattle-keeping. The model compares a model of dependent co-evolution of descent system and cattle-keeping (LD) with a model of cattle and descent system evolving independently of each (LI, see likelihood ratios bottom right), and finds that the model of correlated evolution shown in this flow diagram is a significantly better fit. The thickness of the arrows indicates the rate of change (big arrows and higher rates mean that change happens fast). These results show that matrilineal descent groups who keep cattle are unstable, and rapidly either lose the cattle or the group changes descent system to patriliny. Patrilineal groups with cattle are stable. Thus there is evidence that matrilineal descent systems are associated with lack of cattle. Furthermore, these methods can do more than estimate correlated evolution, as the rates of transition between states shows that it is much more likely that matrilineal groups without cattle first gained cattle and then became matrilineal rather than first becoming patrilineal then gaining cattle (the latter is extremely unlikely to occur). Therefore the likely direction of causation is that a change in the subsistence system changes the social system rather than vice versa.

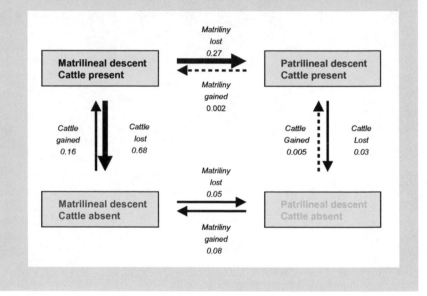

3.2.4 Transmission of Traits from Mother to Daughter Cultures

A key assumption of phylogenetic methods is that the groups under consideration are hierarchically related. The phylogenetic trees that describe the relationships between human populations are best inferred from a neutral trait, such as lexical data. It is then presumed that this is a reasonable model of cultural history. Thus most traits are inherited vertically (along the lineages specified by the branches of trees), rather than transmitted horizontally; but inferring the degree to which individual traits are gained or lost, whether it be spontaneously or by horizontal transmission between groups, and whether this occurs with other traits or independently, provides us with the information we need to distinguish between different models of cultural co-evolution and other evolutionary processes.

There is some confusion in the literature regarding horizontal transmission within and between groups, which have very different implications but are not always clearly distinguished from each other. Horizontal transmission within cultural groups (e.g., social learning from your peers) would be expected if there are such things as 'cultural norms' — indeed it is almost a prerequisite; whereas it is the degree of vertical and horizontal transmission between groups in traits of interest that is relevant to the use of cultural phylogenetic models. Tree-building does require some trait that can be used to infer the main pattern of population history, although some horizontal transmission of traits (sometimes called diffusion) between closely related groups is not very problematic for tree-building (Greenhill, Currie, & Gray 2009). When using phylogenetic comparative methods to seek evidence of correlated evolution, rates of change in a trait on the branches of the tree, be it due to horizontal transmission between groups or spontaneous change, is all part of the data that can be used; large rates of random horizontal transmission can make associations more difficult to detect, but do not invalidate phylogenetic comparative methods, which still function better than non-phylogenetic models when applied to hierarchically related groups (Currie, Greenhill, & Mace, in press). And indeed horizontally transmitted traits, such as subsistence innovations like cattle (as in the example in Box 1), can provide a useful source of cultural variation to use in order to seek evidence for the co-evolution of traits (Mace & Pagel 1994). A high frequency of horizontal transmission of a large number of cultural traits would suggest that trees of lexical data are not necessarily good underlying models for the historical patterns of the cultural history of those norms. However, it is worth noting that even the use of the words 'horizontal' and 'vertical' is predicated on the assumption of an underlying tree-like model, and both would in fact be meaningless terms unless we believed a branching process did indeed underpin our population history, and hence cultural diversification.

Understanding the mode of transmission for different types of cultural variants, and how those variants are exchanged between groups, is an important empirical question, but not explicitly a test of adaptation. Few studies have investigated these processes in a large cross-cultural context. Guglielmino et al. (1995) examined cultural variation in 277 Sub-Saharan African societies coded in the *Ethnographic Atlas* in an attempt to disentangle modes of cultural transmission, while a follow-up by the same group of authors (Hewlett et al. 2002) investigating why African cultures were likely to share traits, added measures of genetic distance to their analyses. In both studies, kinship/family traits were found to be associated with proxies for phylogenetic relatedness, while geographical diffusion explained the distribution of a miscellany of traits with no clear theme, including for example house-building traits and beliefs in high gods. The majority of traits had more than one explanatory model. Ecological correlation was also investigated (even though ecological correlations are rather different, as they were aiming at testing functional adaptation rather than transmission); however, they found that broad ecological categories were not related in any significant way with genetic, linguistic, or cultural similarity. But, as discussed in the previous section, such correlations are not a very good test of ecological adaptation anyway. Neither of these studies controlled for phylogenetic relatedness in a statistical way, and they used broad-scale linguistic classifications across language family boundaries that were at a rather coarse level to address between-society transmission.

3.3 Conclusions

There is overwhelming evidence that wide-ranging aspects of human biology and human behavior can be considered as adaptations to different subsistence systems. Wider environmental and ecological correlates of behavioral and cultural traits are generally best understood as being mediated by differences in subsistence strategies. Thus both cultural and biological factors usually need to be considered together in studies of human evolutionary ecology, combined in specifically defined evolutionary models.

Modeling proximate mechanisms of cultural change is already a well developed field, and some innovative studies are now beginning to test these evolutionary models empirically — although this branch of the field is still in its infancy. Some models predict that generalized social learning mechanisms may cause maladaptive behavior to emerge in some circumstances, but whether such cases are rare or widespread in the real world is not really known.

Models of cultural adaptation to environmental conditions can be subjected to the same or similar tests that behavioral ecologists have used to seek evidence for adaptive behavior in other species. In human ecology, modes of subsistence mediate human interaction with their environment and profoundly influence both human biology, as documented in genetic changes, and human social behavior and cultural norms, such as kinship, marriage, descent, wealth inheritance, and political systems. Phylogenetic comparative methods are proving useful both for studying co-evolutionary hypotheses (be they cultural and/or gene–culture co-evolution), and for estimating ancestral states of prehistoric societies. This form of formal crosscultural comparison is helping to put history back into anthropology, and helping us to understand cultural evolutionary processes at a number of levels.

Acknowledgements I would like to thank Fiona Jordan for a number of comments on this chapter, and the editors for inviting me to contribute.

References

Beaumont MA, Zhang W, Balding DJ (2002) Approximate Bayesian computation in population genetics. Genetics 162:2025–2035

Bersaglieri T, Sabeti PC, Patterson N, Vanderploeg T, Schaffner SF, Drake JA, Rhodes M, Reich DE, Hirschhorn JN (2004) Genetic signatures of strong recent positive selection at the lactase gene. American Journal of Human Genetics 74:1111–1120

Boyd R & Richerson PJ (1985) *Culture and the Evolutionary Process*. University of Chicago Press, Chicago

Currie TE, Greenhill SJ, Mace R (in press) Is horizontal transmission really a problem for phylogenetic comparative methods? A simulation study using continuous cultural traits. Philosophical Transactions of the Royal Society

Currie TE, Mace R (2009) Political complexity predicts the spread of ethnolinguistic groups. Proceedings of the National Academy of Sciences of the United States of America 106:7339–7344

Diamond J (1997) *Guns, Germs and Steel*. Vintage

Flatz G & Rotthauwe H (1973) Lactose, nutrition and natural selection. The Lancet 302:76–77

Fortunato L, Holden C, Mace R (2006) From bridewealth to dowry? A Bayesian estimation of ancestral states of marriage transfers in Indo-European groups. Human Nature. An Interdisciplinary Biosocial Perspective 17:355–376

Gibson MA, Mace R (2006) An energy-saving development initiative increases birth rate and childhood malnutrition in rural Ethiopia. Plos Medicine 3:476–484

Gray RD, Atkinson QD (2003) Language-tree divergence times support the Anatolian theory of Indo-European origin. Nature 426:435–439

Gray RD, Jordan FM (2000) Language trees support the express-train sequence of Austronesian expansion. Nature 405:1052–1055

Greenhill SJ, Currie TE, Gray RD (2009) Does horizontal transmission invalidate cultural phylogenies? Proceedings of the Royal Society B: Biological Sciences 276:2299–2306

Guglielmino CR, Viganotti C, Hewlett B, Cavallisforza LL (1995) Cultural variation in Africa — Role of mechanisms of transmission and adaptation. Proceedings of the National Academy of Sciences of the United States of America 92:7585–7589

Hartung J (1982) Polygyny and the inheritance of wealth. Current Anthropology 23:1–12

Hartung J (1985) Matrilineal inheritance: New theory and analysis. Behavioural and Brain Sciences 8:661–668

Harvey P & Pagel M (1991) *The Comparative Method in Evolutionary Biology*. Oxford University Press, Oxford

Hawks J, Wang ET, Cochran GM, Harpending HC, Moyzis RK (2007) Recent acceleration of human adaptive evolution. Proceedings of the National Academy of Sciences of the United States of America 104:20753–20758

Hewlett BS, de Silvestri A, Guglielmino CR (2002) Semes and genes in Africa. Current Anthropology 43:313–321

Holden C, Mace R (1997) Phylogenetic analysis of the evolution of lactose digestion in adults. Human Biology 69:605–628

Holden CJ, Mace R (2003) Spread of cattle led to the loss of matrilineal descent in Africa: A co-evolutionary analysis. Proceedings of the Royal Society of London Series B: Biological Sciences 270:2425–2433

Ingram CJE, Mulcare CA, Itan Y, Thomas MG, Swallow DM (2009) Lactose digestion and the evolutionary genetics of lactase persistence. Human Genetics 124:579–591

Itan Y, Powell A, Beaumont MA, Burger J, Thomas MG (2009) The origins of lactase persistence in Europe. PLoS Comput Biol 5, e1000491

Johnson AW & Earle T (2000) *The Evolution of Human Societies: From Foraging Group to Agrarian State*. Stanford University Press, Stanford

Jordan FM, Gray RD, Greenhill SJ, Mace R (2009) Matrilocal residence is ancestral in Austronesian societies. Proceedings of the Royal Society B: Biological Sciences 276:1957–1964

Kaplan HS, Hooper PL, Gurven M (2009) The evolutionary and ecological roots of human social organization. Philosophical Transactions of the Royal Society B: Biological Sciences
364:3289–3299

Kendal J, Giraldeau L-A, Laland K (2009) The evolution of social learning rules: Payoff-biased
and frequency-dependent biased transmission. Journal of Theoretical Biology 260:210–219

Laland K, Brown G (2002) *Sense and Nonsense*. Oxford University Press, Oxford

Laland KN, Brown GR (2006) Niche construction, human behavior, and the adaptive-lag hypothesis. Evolutionary Anthropology 15:95–104

Lipo C (2006) *Mapping Our Ancestors: Phylogenetic Approaches in Anthropology and Prehistory*. Wiley

Mace R (1996) Biased parental investment and reproductive success in Gabbra pastoralists. Behavioural Ecology and Sociobiology 38:75–81

Mace R, Holden C, Shennan S (2005) *The Evolution of Cultural Diversity: A Phylogenetic Approach*. Left Coast Press

Mace R, Pagel M (1994) The comparative method in anthropology. Current Anthropology 35:549–
564

Mead S, Whitfield J, Poulter M, Shah P, Uphill J, Campbell T, Al-Dujaily H, Hummerich H,
Beck J, Mein CA, Verzilli C, Whittaker J, Alpers MP, Collinge J (2009) A novel protective
prion protein variant that colocalizes with Kuru exposure. New England Journal of Medicine
361:2056–2065

Murdock GP (1967) *Ethnographic Atlas*. University of Pittsburgh Press, Pittsburgh

Nettle D (2009) Ecological influences on human behavioural diversity: A review of recent findings. Trends in Ecology & Evolution 24:618–624

Odling-Smee J, Laland K, Feldman MW (2003) *Niche Construction: The Neglected Process in
Evolution*. Princeton University Press

Orians GH (1969) On the evolution of mating systems in birds and mammals. American Naturalist
103:589–603

Pagel M (1994) Detecting correlated evolution on phylogenies — A general method for the comparative analysis of discrete characters. Proceedings of the Royal Society of London Series
B: Biological Sciences 255:37–45

Pagel M (1999) Inferring the historical patterns of biological evolution. Nature 401:877–884

Pagel M, Meade A (2005). Bayesian estimation of correlated evolution across cultures: A case
study of marriage systems and wealth transfer at marriage. In *The Evolution of Cultural Diversity: A Phylogenetic Approach* (eds R. Mace, C. Holden, S. Shennan). UCL Press & Left
Coast Press, London

Pagel M, Meade A (2006) Bayesian analysis of correlated evolution of discrete characters by
reversible-jump Markov chain Monte Carlo. American Naturalist 167:808–825

Peng Y, Shi H, Qi X-b, Xiao C-j, Zhong H, Ma R-l, Su B (2010) The ADH1B Arg47His polymorphism in East Asian populations and expansion of rice domestication in history. BMC
Evolutionary Biology 10:15

Perry G, Mace R (2010) Lack of acceptance of evolutionary approaches to human behaviour.
Journal of Evolutionary Psychology 8(2010)2:105–125

Plomin R, DeFries JC, McClearn GE, McGuffin P (2001) *Behavioral Genetics* (4th edn). Worth,
New York

Richerson PJ, Boyd R (2005) *Not by Genes Alone: How Culture Transformed Human Evolution*.
University of Chicago Press, Chicago and London

Segerstrale U (2000) *Defenders of the Truth: The Battle for Science in the Sociobiology Debate
and Beyond*. Oxford University Press, Oxford

Soltis J, Boyd R, Richerson PJ (1995) Can group-functional behaviors evolve by cultural-group
selection — An empirical test. Current Anthropology 36:473–494

Tanaka MM, Kendal JR, Laland K (2009) From traditional medicine to witchcraft: Why medical
treatments are not always efficacious. PlosONE 4:e5192

Chapter 4
Our Selections and Decisions: Inherent Features of the Nervous System?

Frank Rösler

Abstract The chapter summarizes findings on the neuronal bases of decision-making. Taking the phenomenon of selection it will be explained that systems built only from excitatory and inhibitory neuron (populations) have the emergent property of selecting between different alternatives. These considerations suggest that there exists a hierarchical architecture with central selection switches. However, in such a system, functions of selection and decision-making are not localized, but rather emerge from an interaction of several participating networks. These are, on the one hand, networks that process specific input and output representations and, on the other hand, networks that regulate the relative activation/inhibition of the specific input and output networks. These ideas are supported by recent empirical evidence. Moreover, other studies show that rather complex psychological variables, like subjective probability estimates, expected gains and losses, prediction errors, etc., do have biological correlates, i.e., they can be localized in time and space as activation states of neural networks and single cells. These findings suggest that selections and decisions are consequences of an architecture which, seen from a biological perspective, is fully deterministic. However, a transposition of such nomothetic functional principles into the idiographic domain, i.e., using them as elements for comprehensive 'mechanistic' explanations of individual decisions, seems not to be possible because of principle limitations. Therefore, individual decisions will remain predictable by means of probabilistic models alone.

4.1 Introduction

One, if not the most striking feature of living organisms is that they can select between distinct options in order to optimize the outcome of their actions. Seen from

Frank Rösler
Institute of Psychology, University of Potsdam, Karl-Liebknecht-Str. 24/25, 14476 Potsdam OT Golm, Germany, e-mail: froesler@uni-potsdam.de

U.J. Frey et al. (eds.), *Essential Building Blocks of Human Nature*, The Frontiers Collection, DOI 10.1007/978-3-642-13968-0_4, © Springer-Verlag Berlin Heidelberg 2011

the perspective of evolutionary success, this is of extreme advantage. Resources for maintaining the homeostatic equilibrium of an organism are limited and variable in time and space. Therefore, it is most advantageous for survival to actively seek for resources rather than to sit and wait until they accidentally pass by, and it is even more profitable if an organism can, not only actively search, but also select between different options, those that have the highest nutritional value, or those that promise the highest success in reproduction, etc. Active search implies that an organism can make most elementary 'decisions' for sensory processing and locomotion, so there must be basic mechanisms enabling selections between competing input or output options (look left or right, move left or right, etc.). As is well known, such mechanisms are already present in quite primitive organisms, like cubomedusa, a predatory jellyfish that only has ganglia rather than a well-developed nervous system (see Satterlie & Nolen 2001). Therefore, these mechanisms must already emerge if only a limited number of neurons interact.

From a more organism-centered perspective, selection and decision mechanisms are essential too. Organisms are highly restricted in their abilities to process distinct pieces of input information simultaneously, they are unable to perform several distinct movements simultaneously (not only movements that are clearly incompatible), and they are unable to handle more than a limited number of memory representations simultaneously. These input, output, and central processing limitations are vividly present to all of us, both introspectively as well as by observing behavior (for an overview of such cognitive constraints see, e.g., Willingham 2007).

Capacity limitations and trade-offs between different processing resources require mechanisms that regulate the relative dominance of one or the other input source, of memory representations, or of output channels. In cognitive theories such processes are subsumed under the label of selective attention and executive functions (see, e.g., Logan 2004; Fuster 2000). However, examining such theories, it becomes clear that they quite often attribute executive functions to modules that have, more or less explicitly, the status of homunculi. It is said that such modules do the job of information flow regulation, but it is not explained how this is accomplished, how these modules work, or what their intrinsic architecture looks like. This is unsatisfactory. In order to understand human information processing, it is necessary to provide explanations that do not appeal to homunculi, but which describe how such complex processes as sensory selection, memory search, or movement control are enabled by a system that is built up from simple elements (neurons), which interact with each other by nothing more than excitatory and inhibitory connections. In other words, we have to understand the self-organizing characteristics of the system built up from these simple elements and how new, complex functions like selection and executive control actually emerge from the available circuitry.

The fallacy of postulating implicit homunculi is not only a danger for psychological theories. Many papers on the neurophysiology and neuroanatomy of executive functions postulate implicit homunculi as well. For example, there is wide agreement that functions like selection, decision-making, deliberating options in working memory, etc., are in some way related to the frontal, more precisely, to the prefrontal cortex. Nobody can argue against a statement saying that these functions are related to these brain areas or that there is a significant correlation between these two levels of description — neuroanatomy and function. But the situation becomes questionable if research does not go beyond statements like 'structure x of the PFC mediates limited-capacity working memory (WM)', 'structure y controls the parsing of sentences', or 'structure z inhibits impulsive behavior'. This is unsatisfactory because such statements postpone the true explanation, and they also mislead the theoretical analysis in one particular direction, viz., that such functions can be localized within narrowly circumscribed brain areas.

To avoid a misunderstanding here, it is of course very important to map functions onto brain areas — this is the very first step in understanding the architecture of the system. But one should be careful with the words used to describe such an enterprise and the conclusions derived from it. A mapping relationship is a correlation; it does not explain the mechanism! And the presumption that such complex functions can be localized in a simple way by identifying them with narrowly circumscribed brain areas might by completely misleading, as will be outlined below.

In the following, I will cover four topics:

- First, I will briefly discuss some ideas about the way selection mechanisms can be realized with simple elements (neurons) and simple connectivity principles. This will provide a rough blueprint for the architecture of a system that can regulate the relative dominance of activation levels. By so doing I will also point to a central problem that is inherent in the question of how decision mechanisms are implemented in the nervous system, i.e., the convertibility of values (Ramirez & Cabanac 2003; Cabanac 1992). What is the common currency within the nervous system that allows it to rank-order options from highly different content areas? How is the preference regulated between a delicious meal, an opera performance, or spending some time with the children?

- The question of how the architecture might look brings a second aspect into the foreground, namely whether it is possible to identify structures in the CNS that have features predicted by the assumptions made about the architecture. I will present some examples from psychobiological research in which structures that show some of the predicted features are actually identified in the CNS, i.e., which are involved in regulating the relative dominance of activation states.

- Third, I want to show that intervening variables assumed to be relevant for decision-making, e.g., subjective probability or expected gain, are coded within the CNS, i.e., I want to show that rather abstract psychological concepts, which cannot be directly observed but which have to be inferred from experimentally controlled input–output relationships, can be systematically related to physiological measures, e.g., activation changes of larger cell assemblies or of single cells.

- Finally, I will outline some implications and limitations of the approach. In particular, I will discuss the status of such findings with regard to their epistemological value, and how they can contribute to our general understanding of behavior (nomothetic statements) and to the possibility of explaining and predicting individual behavior (idiographic statements).

Last but not least, I will briefly sketch further areas of research that are relevant for the field but which cannot be covered within this chapter.

4.2 Some Thoughts About Selection–Decision Architectures

The nervous system has to solve a selection problem whenever two or more competing subsystems simultaneously seek access for a limited resource. The limited resource can be a common input path, such as the selection of a sector within the visual field, i.e., a decision about where to look; a common output or motor path, such as the selection between incompatible movements, e.g., a decision between a leftward or a rightward movement; or a central processing resource, such as the selection between competing memory representations, between competing homeostatically driven motivations, etc. (McFarland & Sibly 1975).

The saliency of a selected alternative will depend on a variety of extrinsic and intrinsic factors. On the behavioral level, for example, take the options to eat, drink, fight, and flee. The selection of eating vs. drinking will primarily depend on internal homeostatic factors, the current level of energy resources or the current water balance, but of course extrinsic factors like the availability of food and water, the tastiness of a prey, will also be of relevance. On the other hand, for fight and flight, extrinsic factors will be more important than intrinsic factors. If an enemy approaches the organism, it will immediately interrupt feeding or drinking behavior and — depending on some further intrinsic and extrinsic factors — it will either run away or stay and fight.

So there must be a mechanism that accumulates the relevance of all intrinsic and extrinsic causal factors that serve as input to one of the behavioral options. By this means, an overall saliency of each behavioral option will result, and on the basis of this overall saliency, a selection of one or the other option or a switch from one option to another will take place. Such a selection should be terminated when its expression has been successful (a goal has been achieved) or when it has proved to be ineffective, and the possibility of interrupting it must exist, if the competitor for the resource becomes 'stronger'. The selection switching mechanism should act rapidly and to a full extent, i.e., there should be something like a winner-takes-all architecture. This would avoid an oscillation between competing options. On the other hand there should be some hysteresis in the system, i.e., the support of the winner should be enhanced for a short period.

A selection switching mechanism cannot only exist at the highest level on which behavioral strategies are negotiated (see Fig. 4.1). Once a strategy has become dominant due to its overall saliency, e.g., behavior 1 dominates over behavior 2 and behavior 3, it will initiate an action, and a selection between different options must take place once again on the level of actions. If you need to drink, you can either go to the tap or you can fetch a bottle. And, finally, when a particular action has been selected, the relevant movement pattern must be initiated. Here again there is more than one degree of freedom and a selection must take place between the available

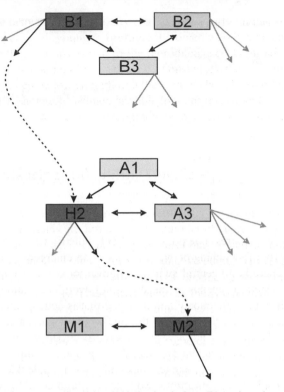

Fig. 4.1 Hierarchy of representational levels: behavioral plans (B), action plans (A), motor programs (M). On each level, a selection of the most dominant representations must take place

movement options, too. A similar hierarchy of selection–switching steps can be assumed for the input side. At a very low level, close to the receptors, there will be competition between feature detectors that are fine-tuned to specific physical stimulus configurations or between distinct locations in space, and at a higher level there will be competition between full percepts, as can be vividly experienced with the Necker cube or other ambiguous drawings, or ambiguous linguistic expressions.

To begin with, it is parsimonious to assume that the selection architecture can be the same on all levels, i.e., that nature has developed one principle to handle the basic problem. Now, how is this achieved with a network of neurons, i.e., with simple elements that can either fire or not fire and that are interconnected by either excitatory or inhibitory synapses?

One rather direct and straightforward approach is a distributed network in which all possibly competing modules are reciprocally inhibitory (see Fig. 4.2). For example, we have a set of input modules, e.g., feature detectors, and the output is fed into the next hierarchical level. An effective selection could be achieved if the inhibition of each module depends on the activation of the other modules. If one module is strongly activated it will automatically inhibit the other modules. This is a positive feedback mechanism: the more one actor is active, the more the others

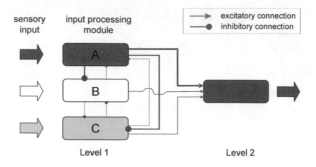

Fig. 4.2 Selection by means of mutually inhibited modules. *Red*: inhibitory connections. *Green*: excitatory connections. *Line thickness* indicates the relative activation level

are inhibited. Let me stress one point here. Such a distributed network model develops the property of switching between inputs as an emergent feature of its wiring architecture. There is no central switch, no governor or control center, no hidden homunculus. Selection depends on the amount of overall activation of each module.

This circuit is fully functional and it is one that is actually implemented in the nervous system, e.g., on the retina of the eye where it guarantees contrast amplification. Nevertheless, it is not a very economical circuit when it comes to many distinct processing modules that are spatially separated. A full mutually interconnected network needs $n(n-1)$ connections and a new module that is added, e.g., by evolution, needs $2n$ new connections. So there are high costs for wiring and when it is in action, there are also high costs in energy expenditure. Moreover, there are some basic anatomical restrictions which make the existence of such a wiring architecture on all levels implausible. Neural interconnections cover on the average only a short distance, a few mm to cm. There are, of course, longer tracti that interconnect whole brain areas with other areas which are separated in space by several centimeters. However, there is no mutual one-to-one interconnectedness, not even within one sensory system — e.g., vision — or all sub-networks of the motor system.

There is another solution to handle the selection problem (see Fig. 4.3). Rather than having mutual connections between each processing module of one level, one can add a centralized selection device. That is, there is a sub-network that receives as information input the net activation levels of each processing module and computes the most dominant activation by means of winner-takes-all mechanisms. This evaluation of the net input is fed back into the input processing modules in such a way that those which are less activated are inhibited. First of all, this architecture is less expensive in wiring and energy expenditure. It needs in total only $2n$ long-distance connections rather than $n(n-1)$, i.e., two for each module, and adding a new module only requires the establishment of 2 further connections rather than $2n$. The device is now modular in the sense that the selection mechanism is decoupled from the more specialized processing modules. Now, one might argue that there is once again a governor or a hidden homunculus. But this is not the case. First, it is possible to design a wiring within the centralized selection module that produces

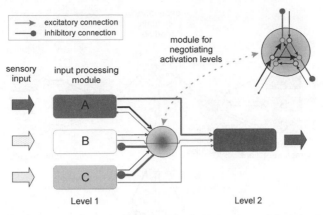

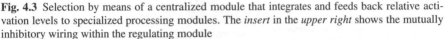

Fig. 4.3 Selection by means of a centralized module that integrates and feeds back relative activation levels to specialized processing modules. The *insert* in the *upper right* shows the mutually inhibitory wiring within the regulating module

the predicted output as an emerging feature once again, i.e., it distributes inhibition depending on the excitatory input. What is needed, of course, is mutual connectivity within the selection module and some connectivity principle that transforms excitation into amplified inhibition. This, however, is not too difficult to build from simple excitatory and inhibitory elements. This solution is anatomically more plausible, because there are only a few long distance connections and the many mutual connections are restricted to a network with a small spatial extension.

Note that this is a distributed network in which none of the elements can do the whole job on its own, and neither is all the information represented at only one particular location. To function properly it needs all the elements — the specific processing modules as well as the specialized switching module — and this means that information is represented in a distributed manner. The selection device, for example, does not have the full representation of the information, but only the overall activation level — it evaluates by its wiring which of the players is crying loudest. On the other hand, the specialized processing modules cannot handle decision/selection on their own.

This wiring architecture has some other advantages. It is not only possible to generalize it easily in a hierarchical manner, but it is also able to solve another problem in a complex system. This problem concerns the convertibility of value information (Cabanac 1992). As mentioned at the beginning, the system has to decide between different behavioral options — eat, drink, mate, or fight. By so doing it does take into account both extrinsic and intrinsic factors. The question is, how can different currencies such as water balance, nutritional value, threat potential, and attractiveness of a mating partner, all be compared to each other?

The solution might be as follows (see Fig. 4.4). This is of course a highly simplified version that does not meet the full complexity of the system. It can only illustrate a working principle. Assume that these switching modules do not only exist for the input devices — within and between sensory modalities — but also that

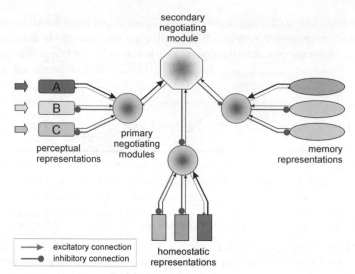

secondary
negotiating
module

A

B

C

perceptual
representations

primary
negotiating
modules

memory
representations

→ excitatory connection
—● inhibitory connection

homeostatic
representations

Fig. 4.4 Generalization of the centralized selection architecture for regulating the relative dominance of specialized processing modules within and between distinct behaviorally relevant domains: sensation, homeostasis, memory. For completeness the motor system has also to be included

a similar selection architecture regulates the relative dominance of memory representations, homeostatic sensors, and motor plans. These localized selection circuits could then be integrated on a higher level by a superordinate selection device which, in principle, has the very same architecture. One advantage is then that the only information that has to be evaluated and compared is the relative activation level of each sub-device.

So far this is rather speculative. However, it could point the way to what we should look for in neuroscience, if we want to disentangle the riddle of how systems reach 'decisions'.

4.3 Switching/Decision/Negotiation Modules

4.3.1 Features of Conflict Negotiation Modules

The previous outline of some construction principles concerning the way the nervous system may have solved the selection problem can provide a route for empirical research. We can start to search for structures in the CNS that have the features one would predict from the layout of the proposed selection architecture. I will briefly outline some of these features and then present some research examples that substantiate corresponding psychobiological relationships, e.g., responses of neuron populations that reflect such features.

An immediate feature that follows from the previous section is that modules like these should be involved if conflicts have to be negotiated, i.e., if the relative dominance of competing plans of action has to be calculated. And as a matter of fact there are neuron populations that respond in such situations and whose activation level reflects the amount of regulation necessary to handle immediate or future input and output conflicts. An example is the anterior cingular cortex (Botvinick et al. 2004; Botvinick et al. 2001).

Apart from conflict negotiations, there is another aspect that is very important for selection and decision situations, viz., the processing of rewards or values. First of all, the system must be able to register the immediate hedonic value of stimuli: are they positive and attractive or negative and threatening? This immediate registration of values will determine whether a plan of action is to be continued or aborted. Immediate evaluation, however, is not enough. Actions are directed towards future stimuli and outcomes. An animal is hungry, therefore it will start moving around to find a prey, and each action and movement has to be evaluated with respect to its prospect value. Therefore, there must be networks or structures in the CNS that calculate expected value and, closely related to this, that calculate outcome probabilities. Here again we already know some areas in the brain that seem to handle these tasks: calculation of expected value and expected outcome probabilities (O'Doherty 2004). Last but not least, the system is adaptive, i.e., it learns from experience, in other words, the system modifies future behavior on the basis of previous outcomes. This needs another mechanism, viz., the calculation of prediction errors which trigger a change in expected values and outcome probabilities (Schultz & Dickinson 2000).

4.3.2 Evidence for Conflict-Negotiating Modules

A very vivid example in which some kind of conflict negotiation seems to take place is the so called Stroop situation, where color words are printed with an incongruent ink, e.g., red, blue, green, yellow. Pronouncing silently the color of the printed words creates a problem. There is obviously some conflict — the printed word is on the tip of the tongue but it has to be inhibited in order to pronounce the color of the ink. This basic task can be presented within a distinct context of expectancies (Carter et al. 2000). A condition can be created under which participants will primarily expect trials with an incongruent word–ink pairing (e.g., 80% incongruent, 20% congruent trials). When going through these trials the participant will, after a while, anticipate the conflict and he or she will strategically suppress or inhibit the activation of the word representation. Under another condition, the word color combinations are such that the participant expects mainly congruent trials (80% congruent, 20% incongruent trials), so he or she will not anticipate much of a conflict and therefore the word representation will be fully activated and dominate over the representation of the color. The two situations make distinct predictions about the experienced and possibly objectively measurable conflict: incongruent trials in a congruent

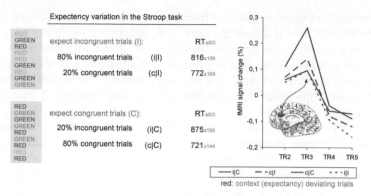

Fig. 4.5 Conflict negotiating effects in the Stroop task, where participants have to read color words printed in a congruent ink (c) or an incongruent ink (i), which are presented in either a congruent (|C) or an incongruent context (|I), i.e., participants can either expect mainly congruent or incongruent trials. *Left*: Sample of the sequence of stimuli under the two conditions. *Middle*: Average response times for congruent and incongruent trials under the two expectancy conditions. *Right*: Average BOLD responses of the anterior cingulate cortex (ACC) for the four events. The ellipse on the medial wall of the right hemisphere shows the approximate location of the ACC. Data from Carter et al. (2000)

context will create more conflict than incongruent trials in an incongruent context, and congruent trials in an incongruent context will most likely be more difficult to handle than congruent trials in a congruent context. This is exactly what one can see in the average response times of a group of 12 young healthy participants (see Fig. 4.5, top left).

Carter et al. (2000) used this design and simultaneously measured the blood oxygenation level dependent (BOLD) effect by means of functional magnetic resonance imaging (fMRI). This is a signal that reflects the oxygen consumption of neural tissue. With this method it is possible to see which regions of the brain need more or less energy (oxygen supply) in a particular processing epoch (Huettel et al. 2004). Blood flow changes are very precisely tuned and highly localized in the brain due to the capillary structure of the supplying vessels. When neurons fire they need energy, and this energy is supplied by metabolic processes. These metabolic processes need oxygen, and due to the demand-dependent regulation, oxygen is delivered where it is needed. The observed signal is the same as the change of your skin when your face is blushing — it is a localized change of blood flow.

Returning to the study, Carter et al. (2000) looked at those brain areas that were most active in the conflict situation created by the Stroop test. In fact, they did not do a blind search in this study, because they already knew from previous work that a likely candidate is the anterior cingulate cortex (ACC) of the brain. The striking finding of this study is the graded changes of the activity within the ACC (Fig. 4.5). The maximum BOLD signal appears in the situation in which most conflict is expected, both from introspection and from response times, i.e., when congruent items are expected but an incongruent trial has to be processed (i|C). The second largest activation appears in the conflict situation, when congruent trials have to be handled in an incongruent environment, i.e., when word processing is strategically suppressed, but when a match between word and color exists (c|I). The graph shows that it is not the incongruency as such that activates the ACC, nor is it the probability of the specific trial, rather it is a context-dependent conflict situation. Thus, the ACC seems to be one structure we are after, that is, it seems to be a network that regulates conflicting activation levels.

Of course, this cannot be concluded from a single study. But there are many studies with similar designs that showed equivalent effects (see Botvinick et al. 2004 for a summary). These data suggest that with some spatial variation — due to methodological differences — very distinct paradigms resulted in highly similar activations of the ACC, i.e., if preponderant reaction tendencies have to be inhibited, if a selection from alternatives with equal action probability has to be performed, or if a discrepancy between action plan and action is encountered. It makes no difference whether the response conflict is vocal, manual, or oculomotor. This shows that there is an independence from task-specific variables. Moreover, it can also be shown that the effects observed in the ACC can be simulated in a computer model with such a negotiating module, which feeds back to the specific processing modules and in this way amplifies or inhibits the relative level of activation (Botvinick et al. 2001).

Other research has shown that t-switching or decision modules can also be found at other brain locations. Although they all seem to be based on the same functional principle, in that they integrate relative activation levels of specialized processing modules and feed these back via inhibitory connections in order to amplify relevant and attenuate irrelevant information, they nevertheless show functional specializations. Thalamic nuclei, in particular the nucl. reticularis thalami, regulate input-related selective attention (Yingling & Skinner 1977; LaBerge 1995), the anterior cingulum is involved in response-related dominance regulation (Botvinick et al. 2004; Rushworth et al. 2007), and some areas of the prefrontal cortex regulate the relative dominance of representations held in working memory (Badre & Wagner 2007; Constantinidis et al. 2001). All in all, these conflict regulation modules are not the 'super-switch' that negotiates between all domains as sketched in Fig. 4.4. Rather they seem to be more domain-specific in their regulation capabilities.

4.4 Representation of Decision-Related Intervening Variables

As mentioned earlier, another feature of decision modules must be the coding of expected reward and outcome probabilities. During the last few years, several groups have started to explore whether and how decision-related intervening variables are coded in the firing rate of individual neurons. Of course, this work is carried out on animals, mostly macaque monkeys, which have implanted electrodes to register task-related signals (Gold & Shadlen 2001; Romo & Salinas 2003; Glimcher 2003). Another approach tries to establish similar psychobiological relationships in humans by using brain-imaging tools, e.g., fMRI (see, for example, Kringelbach

2005; Rushworth et al. 2004; Tanaka et al. 2004; Heekeren et al. 2003; Knutson & Cooper 2005; Daw et al. 2006). I will present an example from each line of research.

4.4.1 BOLD Responses in Humans

Imagine you have to choose one out of four one-armed bandits. With each slot machine you can win a certain number of points, later to be converted into money. If you know nothing about the situation, you will have to find out which slot machine is the best in order to maximize your gains. What you do not know, but what you will learn by trial and error, is that the payoff of each machine will vary over time. So a machine that provides high gains at the beginning may become a less attractive choice later on, because the payoffs are then smaller compared to the other machines. The variation of payoffs over time is realized by a decaying random-walk process.

Now, what is the best strategy to cope with this situation? First you will more or less randomly choose one or the other machine. After a while you will find out that machine A provides the best payoff. It would be reasonable to stay with this machine, to exploit it. But after a while the payoffs will decrease due to the random walk diffusion process. After noticing this change it would be wise to explore the other options again. So a rational strategy would be to switch between exploitation and exploration. By using a time series model, a Kalman filter (Anderson & Moore 1979), it is possible to estimate the expected gain, the prediction error, and the choice probabilities of each option from the behavior of each participant at each trial.

Daw et al. (2006) registered the BOLD signal of the brain by means of functional magnetic resonance imaging and asked whether narrowly localized brain regions can be related to parameters estimated from the data by means of an adaptive learning model. These were expected gain, prediction error, and choice probabilities. Moreover, they wanted to know whether two hypothesized strategies, exploration vs. exploitation, are related to an activation of distinct areas. To summarize the findings of this and many other studies with similar experimental setups, it is quite clear that such intervening variables can be related to distinct brain activation patterns. The immediate reward value activates the orbito-frontal cortex, while the expected reward value activates the ventro-medial prefrontal cortex. The difference between these two intervening variables, the prediction error in a given trial, is computed elsewhere, viz., deeper in the brain, within parts of the basal ganglia — the nucl. accumbens and the nucl. caudatus (see Fig. 4.6). This accords with many other findings from neurochemistry that show an increased dopamine signal in these structures whenever a large prediction error has to be assumed (Schultz 2000; Schultz 2002).

There is another finding that is particularly interesting. With adequate estimation procedures and appropriate experimental manipulations, it is possible to estimate whether a participant takes into account rewards in the future or those more immediately available. The prediction errors based on these different time perspectives

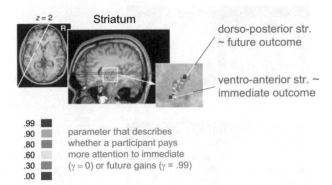

Fig. 4.6 Correlations of fMRI activations in the striatum with reward prediction error related to either immediate or future gains. *Blue* means that the participant pays attention primarily to future expected reward and that errors in this prediction activate the dorso-posterior part of the striatum. In contrast, *red* means that the participant pays attention to immediate rewards. If she errs on these, the ventral-anterior part of the striatum becomes activated. Data reprinted with permission from Macmillan Publishers Ltd, Nature Neuroscience, from Tanaka et al. (2004), Fig. 6

seem to be processed at different locations in the striatum. And as shown in Fig. 4.6, there are neuron populations located within a few millimeters of brain tissue in the striatum whose activation level covaries with errors on immediate and/or future expected reward values.

Last but not least, there is a further important finding in the study by Daw et al. (2006). They were able to show that an evaluative, rational strategy of exploration activates the frontal pole, or as labeled by others, the dorso-lateral prefrontal cortex. What does 'exploration' mean? It means a shift away from the most attractive option to a less attractive option. It means in some sense an inhibition of an otherwise strong response tendency. However, this more hesitant, deliberating selection process does not exclusively involve the narrowly circumscribed brain area just mentioned. Rather, it is a network that is involved, as predicted by the more general speculations about the selection architecture given above (see Fig. 4.7). While the prefrontal cortex is activated in this situation, other areas of the brain are activated too, in particular more posterior areas that are functionally related to mental imaging processes. So, again, despite the attempts to localize certain biopsychological relationships, it has to be concluded that such a mapping is not one-to-one. There is no central controller or homunculus-type brain region that does the job. Rather, there is a distributed network in which relative activation levels are negotiated and regulated by feedback loops.

It is interesting to notice that, if the dorso-lateral prefrontal cortex is knocked out in a game-playing situation for a couple of minutes by transcranial magnetic stimulation (TMS), the participants do show more impulsive behavior. Among others, they then accept unfair assets with less hesitation (Knoch et al. 2006). As a matter of fact, such a less critical, more impulsive behavior also holds for patients with a lesion in this very brain area, due to an accident or a stroke. Thus, the dorso-lateral

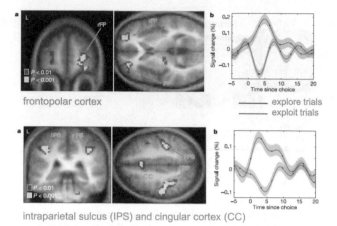

frontopolar cortex

intraparietal sulcus (IPS) and cingular cortex (CC)

Fig. 4.7 Exploration-related BOLD effects in the frontopolar cortex (*top*) and intraparietal sulcus (*bottom*). In both cases, the contrast between explore and exploit trials is shown, as estimated from the behavior time series. *Right*: Hemodynamic response functions revealing a significant amplitude difference between the two conditions. Data reprinted with permission from Macmillan Publishers Ltd, Nature, from Daw et al. (2006), Figs. 3 and 4

prefrontal cortex is essential for regulating behavioral dominance but, as revealed by brain imaging, it is part of a widespread network whose intrinsic wiring is not yet fully understood. It might be based on the principles of regulating relative activation levels by means of the architecture presented, but further studies are needed to substantiate this.

4.4.2 Single Cell Responses in Animals

Although BOLD signals provide insight into where certain processing steps are performed in the brain and which areas might work together in a particular experimental situation, one has to keep in mind that the BOLD response is an indirect measure of neural activity alone. It reflects the energy or oxygen expenditure of several thousand neurons and it does so quite indirectly, with a considerable delay (~ 2 s). Moreover, the BOLD signal is blind to the functional properties of the activation, whether it is excitatory or inhibitory, or whether it is primarily due to afferent, centrally connecting, or efferent cell populations. Thus, fMRI studies cannot reveal the wiring architecture on the micro-level of individual neurons.

Therefore, such experiments need to be transferred into the animal lab, where we may ask whether equivalent psycho-physiological relationships can also be detected on the level of individual neurons, i.e., whether the firing rate of a neuron can be related to intervening variables like expected gain or expected probabilities. This is an even greater challenge than studies with humans, because one requires a task that a monkey can understand and a motor response that indicates the monkey's decision, and one has to manipulate the variables of interest in such a way that intervening variables, like subjective outcome and estimated probabilities, are actually influenced systematically. A simple task often used for this enterprise, and easy to train a monkey to do, employs controlled saccadic eye movements towards a specified target. A correct eye movement (correctness defined by the experimenter on the basis of the design) will be reinforced by a raisin or a drop of fruit juice. Thus, the monkey first sees a cue, upon which it has to decide where to look. Then, after an imperative stimulus, the eyes have to be moved towards the target. If the movement is correct, reinforcement will be given.

In such a setup, the firing rates of neurons can be recorded in the cortex, in particular in areas known to be functionally related to visual processing and eye movement control. Neurons have a variable firing rate and the frequency of firing characterizes the functional features of a neuron. For example, in the visual cortex, some neurons fire preferentially for stimuli that appear in certain locations of the visual field and which have a specific orientation. In the motor cortex, the firing rate of some neurons depends on the direction of a movement. These are straightforward physical-physiological relationships, e.g., if the monkey looks upward, a particular neuron fires, whereas if it looks downward, the same neuron does not fire, but then another neuron, next to the first, may show the opposite firing behavior, a maximum with a downward movement, and no firing with an upward movement. In short, such neurons can be labeled according to their dominant firing response as up, down, left, or right neurons. However, these neurons do not only indicate the direction of a saccadic eye movement.

As a matter of fact, it could be shown that the firing rate of direction-specific movement neurons is modulated by the expected gain associated with a certain movement direction (see Fig. 4.8) (Platt & Glimcher 1999). For example, a neuron that is tuned to an upward movement will show a higher firing rate if the expected gain for an upward movement is large compared to the situation when the expected gain for the same movement is small. Several aspects of this result are noteworthy. First, the modulation of the firing rate precedes the movement, i.e., it precedes overt behavior. This means that the motivational variable 'expected gain' has a direct influence on a motor neuron and its activity. Second, the correlation between firing rate and expected gain — shown by the regression lines in Fig. 4.8C — disappears with the progression of the trial. When the monkey actually moves its eyes, the firing rate is driven more by the direction of the movement than by the expected gain variable. This means that the neuron must be integrated into different circuits. It is transiently bound to the expectancy–reward circuit, and then it is transiently bound to the movement circuit. However, most important is the fact that such a complex concept as expected gain is reflected at all in the firing rate of individual neurons.

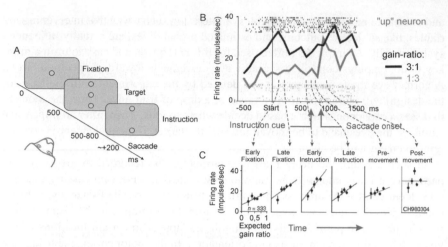

Fig. 4.8 (**A**) Design. A monkey has to fixate a central LED. After 500 ms, two eccentric LEDs appear that indicate movement fields of neurons located in the parietal cortex. A saccadic eye movement towards these locations evokes a maximum firing rate in the related neurons, e.g., in an up neuron or a down neuron. (**B**) Firing rates of a neuron in the parietal cortex of a macaque monkey that is tuned to an upward movement of the eyes. The *two graphs* show the average firing rate over a trial in which the monkey has to initiate a gaze shift towards a target, but not earlier than an instruction cue is given later in the trial. The *black curve* shows the firing rate if the gain ratio is .75, i.e., if the monkey receives 3 times as much juice for an upward as for a downward movement, while the *gray curve* shows the firing rate for the opposite gain ratio .25. (**C**) Correlations between firing rate and expected gain during distinct epochs of a trial. As can be seen, the correlation is high after the target has been processed and before the motor response has actually been executed. Later on, the correlation drops again, i.e., then the neuron is mainly driven by the response direction, rather than by the expected gain. Data reprinted with permission from Macmillan Publishers Ltd, Nature, from Platt & Glimcher (1999), Fig. 3

4.5 Summary and Some Implications

4.5.1 Distributed Networks and Complex Interactions

With sophisticated experimentation and appropriate measurement tools it is possible to naturalize highly abstract psychological concepts that are derived from decision-theoretic models and from introspection. Neurons and neural networks can be identified whose activation pattern correlates closely with intervening variables like expected gain or subjective probability, i.e., variables that have to be inferred from the observables — the experimental manipulations, the measured behavior, and introspective reports.

Although such neurobiological correlates can be localized in the sense that functional aspects are related to narrowly circumscribed brain areas, one should not be misled by this into concluding that a function as such is fully represented by a particular brain site. A function mediating selection or decision behavior or an intervening variable essential for this is never fully localized in one, narrowly defined cell assembly or even to a single neuron. Rather, such a function emerges out of the interaction of neurons in a widely distributed network, and a particular behavioral output is always the result of highly complex interactions between many, if not all parts of the brain. Moreover, each type of behavior also involves many, evolutionarily distinct levels, i.e., it is tied neither to the cortex nor to subcortical entities alone. In other words, it is never restricted to just one level of the evolution-dependent hierarchy of the nervous system. It is the interaction that counts, and it is the ensemble of all participating cell assemblies which constitute a psychological function. For example, if input from a lower level is missing, the switching module on a higher level, as outlined in Fig. 4.2, cannot do its work properly, because the balance of excitatory and inhibitory activations will be disturbed. This is impressively demonstrated by some paradoxical lesion effects. If a higher level is destroyed and a function is lost, the function can be partially re-established by additionally lesioning a lower-level structure. A convincing example is the so-called Sprague effect (Wallace et al. 1990).

No doubt the architecture of and the interactions within the CNS are complex, notwithstanding the fact that the whole system is built from very simple elements with strikingly limited functional features, viz., neurons that excite or inhibit other neurons. However, the basic connectivity principles outlined above show that, in principle, complex functions like selection and decision-making can be realized in a straightforward manner by combining these building blocks. Moreover, data from neuroscience prove that structures and their functions suggested by such models can actually be identified within the nervous system.

4.5.2 Nomothetic vs. Idiographic Descriptions

What is the scientific status of such findings? What do they imply, for example, for the explanation of the individual or social behavior of humans?

First of all, it has to be admitted that experiments in neuroscience as presented above do only scratch the surface of the problem. The experiments bring about fascinating results, no doubt, and sometimes they may even be frightening, because we have a vague feeling of some far-reaching implications. Nevertheless, they only provide insights on the level of principles, and do not — at least for the time being — allow exact predictions of individual behavior. This is often not correctly perceived either by neuroscientists or by their critics. Learning about psychobiological relationships, knowing that even highly complex psychological variables can be naturalized and that our behavior rests on physiology, for which the causal laws of macrophysics hold, leads many to the conclusion that physical determination is a simple, straightforward thing, and that our brain is a machine that can easily be understood, like a car, a wristwatch, or, to make it a bit more complicated, a personal computer.

But here is a warning: one should not underestimate the complexity of the system! To illustrate this, just consider some numbers. The total number of neurons in our brain amounts to roughly 10^{12}, and on average, each neuron has between 10 000 and 15 000 connections with other neurons. Thus, in total there are about 10^{16} synapses, i.e., contact points that determine a momentary state and the next state shift of the system. Of course, there is some redundancy in the system, that is, clusters of neurons that show equivalent behavior. Nevertheless, the number of combinations that lead to a specific momentary state is immense and can hardly be grasped by our imagination. The number of stars in our galaxy, the Milky Way, amounts only to 2×10^{11}. Thus, the point I want to make is that we are able to describe some general functional principles of the system (nomothetic descriptions), just as a physicist can do this for atoms, molecules, and assemblies of such, but we cannot lay out the full deterministic chain of causality that will lead to individual behavior (idiographic descriptions). This is impossible for the time being, and it will most likely remain impossible in the future, simply because we cannot simultaneously monitor all the synaptic contacts that define the state. Likewise, it is impossible to capture the full chain of causality from the motions of individual molecules to the movement of a whole car.

The situation is even more complicated than with any physical system. Living organisms are adaptive, they learn, and on the physiological–anatomical level this means that the microstructure of the system changes as a consequence of the previous system state. Likewise, the environment also changes due to the behavior of the organism. This means that we have a highly interdependent network of internal and external causes in which activity states show nonlinear dynamics (see Baltes et al. 2006a).

Let us assume that behavior b_t at time t and its related mental states are mapped in a one-to-one relation onto underlying transient states a_t of neural activity (these are the transition states of the 10^{16} synaptic contacts). This activity pattern depends on one side on the neural circuitry (y_t) that has developed from genetic predispositions and from the individual learning history, while on the other side, it depends on the currently encountered stimulus configuration in the environment (x_t) (see Fig. 4.9). This configuration not only comprises the stimuli that we might process consciously, but the total set includes much more, and in addition, not all the stimuli are consciously processed. So the current activity pattern in 10^{16} synaptic connectivity points is the convolution of two vectors which both have an extremely high dimensionality, as expressed by

$$b_t \equiv a_t = f(x_t \times y_t) . \tag{4.1}$$

This is the situation at time t. What happens at time $t + 1$? Here again we have the convolution of the microstructure of the brain and the new stimulus configuration (4.2), but, and this is the important point, both are a function of the preceding states (4.3) and (4.4). The new neural network depends on its previous state, but also the previous activity pattern.

Each and every activity in the nervous system results in changes of the circuitry. This is the biological basis of all learning phenomena (Buonomanon & Merzenich 1998; Kolb & Whishaw 1998; Fanselow & Poulos 2005; Ming & Song 2005; Quartz

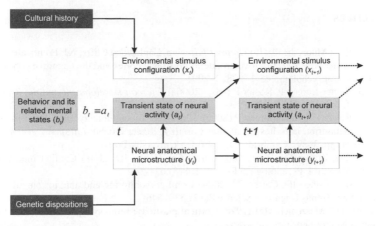

Fig. 4.9 Formalization of the co-construction hypothesis, illustrating the interaction between continuous changes in the internal microstructure and functional states of an organism and the continuous changes in the environment. Adapted from Baltes et al. (2006b)

& Sejnowski 1997). On the other hand, as the organism interacts with the environment, the new stimulus configuration depends on the previous stimulus configuration plus the induced changes in the organism. Bringing this together means inserting (4.3) and (4.4) into (4.2). This reveals immediately that there exists a convolution of two recursive functions (4.5):

$$a_{t+1} = f(x_{t+1} \times y_{t+1}) , \qquad (4.2)$$

$$x_{t+1} = g(a_t \times x_t) , \qquad (4.3)$$

$$y_{t+1} = h(a_t \times y_t) , \qquad (4.4)$$

$$a_{t+1} = f\left[g(a_t \times x_t) \times h(a_t \times y_t)\right] . \qquad (4.5)$$

This results in a highly complex system that is nonlinear and dynamic. Predictions in such a system cannot be easily captured by deterministic laws, by the cause and effect statements we are so used to.

Consequently, as far as the nervous system and human or animal behavior is concerned, we can only make probabilistic predictions. For the time being, such probabilistic predictions relating the behavioral level to the biological level are pretty weak. They may become more precise in the future, but nevertheless, a probabilistic prediction can never be deterministic and causal in the strict sense. They always imply the possibility of prediction errors, a caveat that is often not addressed, e.g., by researchers claiming that the criminal dispositions of an individual can be predicted on the basis of biological signatures (Roth 2004; Yechiam et al. 2008).

Acknowledgements The chapter is based on a lecture given at the Institute for Advanced Study, Berlin, during the academic year 2006/2007. The support of the Center for Advanced Study is gratefully acknowledged.

References

Anderson BDO, Moore JB (1979) *Optimal Filtering*. Englewood Cliffs, NJ: Prentice Hall
Badre D, Wagner AD (2007) Left ventrolateral prefrontal cortex and the cognitive control of memory. Neuropsychologia 45(13):2883–2901
Baltes PB, Reuter-Lorenz P, Rösler F (eds) (2006a) *Lifespan Development and the Brain*. Cambridge: Cambridge University Press
Baltes PB, Rösler F, Reuter-Lorenz P (2006b) Prologue: Biocultural co-constructivism as a theoretical metascript. In Baltes PB, Reuter-Lorenz P, Rösler F (eds): *Lifespan Development and the Brain* (pp. 3–39). Cambridge: Cambridge University Press
Botvinick MM, Braver TS, Barch DM, Carter CS, Cohen JD (2001) Conflict monitoring and cognitive control. Psychological Review 108(3):624–652
Botvinick MM, Cohen JD, Carter CS (2004) Conflict monitoring and anterior cingulate cortex: An update. Trends Cognitive Science 8(12):539–546
Buonomano DV, Merzenich MM (1998) Cortical plasticity: From synapses to maps. Annual Review of Neuroscience 21:149–186
Cabanac M (1992) Pleasure: The common currency. Journal of Theoretical Biology 155(2):173–200
Carter CS, McDonald AM, Botvinick M, Ross LL, Stenger VA, Noll D, Cohen JD (2000) Parsing executive processes: Strategic vs. evaluative functions of the anterior cingulate cortex. Proceedings of the National Academy of Science USA 97:1944–1948
Constantinidis C, Franowicz MN, Goldman-Rakic PS (2001) The sensory nature of mnemonic representation in the primate prefrontal cortex. Nature Neuroscience 4:311–316
Daw ND, O'Doherty JP, Dayan P, Seymour B, Dolan RJ (2006) Cortical substrates for exploratory decisions in humans. Nature 441(7095):876–879
Fanselow MS, Poulos AM (2005) The neuroscience of mammalian associative learning. Annual Review of Psychology 56:207–234
Fuster JM (2000) Executive frontal functions. Experimental Brain Research 133:66–70
Glimcher PW (2003) The neurobiology of visual–saccadic decision-making. Annual Review of Neuroscience 26:133–179
Gold JI, Shadlen MN (2001) Neural computations that underlie decisions about sensory stimuli. Trends in Cognitive Science 5(1):10–16
Heekeren HR, Wartenburger I, Schmidt H, Schwintowski HP, Villringer A (2003) An fMRI study of simple ethical decision-making. Neuroreport for Rapid Communication of Neuroscience Research 14(9):1215–1219
Huettel SA, Song AW, McCarthy G (eds) (2004) *Functional Magnetic Resonance Imaging*. Sunderland, MA, USA: Sinauer Ass. Inc.
Knoch D, Pascual-Leone A, Meyer K, Treyer V, Fehr E (2006) Diminishing reciprocal fairness by disrupting the right prefrontal cortex. Science 314(5800):829–832
Knutson B, Cooper JC (2005) Functional magnetic resonance imaging of reward prediction. Current Opinion in Neurology 18(4):411–417
Kolb B, Whishaw IQ (1998) Brain plasticity and behavior. Annual Review of Psychology 49:43–64
Kringelbach ML (2005) The human orbitofrontal cortex: Linking reward to hedonic experience. Nat Rev Neurosci 6(9):691–702
LaBerge D (1995) Computational and anatomical models of selective attention in object identification. In Gazzaniga MS (ed): *The Cognitive Neurosciences* (pp. 649–663). Cambridge, MA: MIT Press
Logan GD (2004) Cumulative progress in formal theories of attention. Annual Review of Psychology 55:207–234
McFarland DJ, Sibly RM (1975) The behavioural final common path. Philosophical Transactions of the Royal Society of London Series B: Biological Sciences 270(907):265–293
Ming Gl, Song H (2005) Adult neurogenesis in the mammalian central nervous system. Annual Review of Neuroscience 28:223–250

O'Doherty JP (2004) Reward representations and reward-related learning in the human brain: Insights from neuroimaging. Current Opinion in Neurobiology 14(6):769–776

Platt ML, Glimcher PW (1999) Neural correlates of decision variables in parietal cortex. Nature 400(6741):233–238

Quartz S, Sejnowski TJ (1997) The neural basis of cognitive development: A constructivist manifesto. Behavioral and Brain Sciences 20(4):537–596

Ramirez JM, Cabanac M (2003) Pleasure, the common currency of emotions. Annals of the New York Academy of Sciences 1000:293–5

Romo R, Salinas E (2003) Flutter discrimination: Neural codes, perception, memory and decision-making. Nat Rev Neurosci 4(3):203–218

Roth G (2004) Freier Wille, Verantwortlichkeit und Schuld. In Berlin-Brandenburgische Akademie der Wissenschaften (ed), *Zur Freiheit des Willens* (pp. 63–70). Berlin: Akademie Verlag

Rushworth MF, Behrens TE, Rudebeck PH, Walton ME (2007) Contrasting roles for cingulate and orbitofrontal cortex in decisions and social behaviour. Trends in Cognitive Science 11(4):168–176

Rushworth MF, Walton ME, Kennerley SW, Bannerman DM (2004) Action sets and decisions in the medial frontal cortex. Trends in Cognitive Science 8(9):410–417

Satterlie RA, Nolen TG (2001) Why do cubomedusae have only four swim pacemakers? Journal of Experimental Biology 204(Pt 8):1413–1419

Schultz W (2000) Multiple reward signals in the brain. Nat Rev Neurosci 1(3):199–207

Schultz W (2002) Getting formal with dopamine and reward. Neuron 36(2):241–263

Schultz W, Dickinson A (2000) Neuronal coding of prediction errors. Annual Review of Neuroscience 23:473–500

Tanaka SC, Doya K, Okada G, Ueda K, Okamoto Y, Yamawaki S (2004) Prediction of immediate and future rewards differentially recruits cortico-basal ganglia loops. Nature Neuroscience 7(8):887–893

Wallace SF, Rosenquist AC, Sprague JM (1990) Ibotenic acid lesions of the lateral substantia nigra restore visual orientation behavior in the hemianopic cat. Journal of Comparative Neurology 296(2):222–252

Willingham DT (2007) *Cognition: The Thinking Animal.* (3rd edn) New Jersey: Pearson Education Inc

Yechiam E, Kanz JE, Bechara A, Stout JC, Busemeyer JR, Altmaier EM, Paulsen JS (2008) Neurocognitive deficits related to poor decision-making in people behind bars. Psychon Bull Rev 15(1):44–51

Yingling CD, Skinner JE (1977) Gating of thamalic input to cerebral cortex by nucleus reticularis thalami. In Desmedt JE (ed): *Progress in Clinical Neurophysiology* (Vol. 1, pp. 70–96). Basel: Karge

Chapter 5
Our Gods: Variation in Supernatural Minds

Benjamin G. Purzycki and Richard Sosis

Abstract In this chapter we examine variation in the contents of supernatural minds across cultures and the social correlates of this variation. We first provide a sketch of how humans are capable of representing supernatural minds and emphasize the significance of the types of knowledge attributed to supernatural agents. We then argue that the contents of supernatural minds as represented cross-culturally will primarily rest on or between two poles: knowledge of people's moral behavior and knowledge of people's ritualized costly behavior. Communities which endorse omniscient supernatural agents that are highly concerned with moral behavior will emphasize the importance of shared beliefs (cultural consensus), whereas communities which possess supernatural agents with limited social knowledge who are concerned with ritual actions will emphasize shared behavioral patterns (social consensus). We conclude with a brief discussion about the contexts in which these patterns occur.

5.1 Introduction

Wilson remarked that "religions are like other human institutions in that they evolve in directions that enhance the welfare of the practitioners" (Wilson 1978, p. 182). Here, we attempt to detail one such way in which religious traditions change in order to maximize the benefits reaped from participation. We are probably the only species on Earth capable of pondering the existence of 'spiritual agencies', as Darwin considered religion (Darwin 1871/2004, p. 117), so we begin our discussion with the cognitive mechanisms required to conceive of these agents. However, we do not limit our representations of these agents to their minds. We also concern ourselves with their movements and mental states. More specifically, we commit ourselves to

Benjamin G. Purzycki
Department of Anthropology, University of Connecticut, 354 Mansfield Road, Storrs, CT, USA,
e-mail: benjamin.purzycki@uconn.edu

Richard Sosis
Department of Anthropology, University of Connecticut, 354 Mansfield Road, Storrs, CT, USA,
e-mail: richard.sosis@uconn.edu

U.J. Frey et al. (eds.), *Essential Building Blocks of Human Nature*, The Frontiers
Collection, DOI 10.1007/978-3-642-13968-0_5, © Springer-Verlag Berlin Heidelberg 2011

these agents and consequently either refrain from or perform a host of behaviors to appease and please them. These commitments to supernatural agents form the core of the human religious system. This system evolved to overcome significant adaptive problems of human sociality, particularly problems of coordination and cooperation (Alcorta & Sosis 2005; Purzycki & Sosis 2009; Rappaport 1999; Sosis 2009a).

Here, we focus our attention on the ability to attribute mental states to supernatural agents, the way knowledge attributed to supernatural agents motivates behavior, and why such knowledge and concomitant behaviors vary across communities in specific ways. To summarize our argument, supernatural agents vary widely in form across cultures: some are people-like, others are animal-like, and some are conceived of as bodiless forces. The knowledge attributed to these varying forms seems to be constrained between two poles. Some supernatural agents are believed to be omniscient, but in fact these agents are primarily concerned with human moral behavior. Other agents are believed to possess limited social knowledge and are not concerned with human morality, but rather are concerned with ritual performance. Societies develop along these divergent trajectories depending on the ability of communities to monitor behavior.

5.2 Representing Supernatural Minds

5.2.1 The Mindreading System and Attributed Domains

Supernatural agent concepts are found in every human society. From gods and ghosts to ancestor and animal spirits, supernatural agent beliefs exhibit not only many essential similarities but also remarkable differences across cultures. The similarities frequently suggest pan-human cognitive biases, whereas the differences are often considered to be merely cultural byproducts of evolved cognitive mechanisms (Atran 2002; Atran & Norenzayan 2004; Boyer 2001; Kirkpatrick 2006, 2008). One such bias is the ability to explain events and rationalize behavior in terms of supernatural agents. We begin our chapter with a sketch of the human attribution of mental states to other entities and objects in the world. In contrast to the dominant view in the cognitive science of religion (Boyer & Bergstrom 2008), we have argued that, in the context of religion, such attributions show evidence of functional design (Purzycki & Sosis 2010; Sosis 2009a). Here we further develop this argument by exploring the variation in the types of knowledge attributed to supernatural minds around the world.

Detecting mental states has often been characterized as an exclusively human characteristic, although there are indications that other primates have the ability to do so as well (Call & Tomasello 2008). This ability to detect and represent mental states has come to be known as 'theory of mind' or ToM (Premack & Woodruff 1978). Even though we ultimately lack direct, solid evidence of each others' minds — let alone their contents — we nevertheless cannot help but attribute internal motivational states to animate entities. Baron-Cohen (1997) characterizes the mindreading complex as composed of subsystems that infer internal motivational states driving the observable behavior of other animate entities. One such subsystem is the intentionality detector, which "interprets motion stimuli in terms of the primitive volitional mental states of goal and desire" (Baron-Cohen 1997, p. 32). The intentionality detector "is activated whenever there is any perceptual input that might identify something as an agent. [...] This could be anything with [apparently] self-propelled motion. Thus, a person, a butterfly, a billiard ball, a cat, a cloud, a hand, or a unicorn would [activate this mechanism]" (Baron-Cohen 1997, p. 33; see Gelman et al. 1996). Baron-Cohen (1997) suggests that the intentionality detector's "value lies in its generality of application: it will interpret almost anything with self-propelled motion, or anything that makes a non-random sound, as a query agent with goals and desires" (Baron-Cohen 1997, p. 34).

Barrett and Keil (1996; Barrett 2004) argue that we have a mental mechanism, similar to Baron-Cohen's intentionality detector, which interprets objects and events in terms of agency. They refer to this mechanism as the hyperactive agency detection device (HADD). HADD is hyperactive insofar as it attributes agency even to agentless events and things such as rustling bushes, moving dots on a computer screen, surprising events, and most importantly for our discussion, the gods and spirits of the world's religious traditions. HADD can of course be overridden by *post hoc* conscious reflection. What makes Barrett's account significant for religious cognition is that the hyperactivity of this device triggers the ToM to explain events and other phenomena in agentive terms. But the specific *form* of the agent causing a mysterious event systematically varies cross-culturally (see below). The subsystems of the mindreading system, including HADD or the intentionality detector, share most of the features of Fodor's conservative definition of cognitive modules (Baron-Cohen 1995, pp. 57–58; Fodor 1983, 2000). As input-only mechanisms with operations we cannot consciously manipulate (i.e., it is difficult *not* to detect internal states upon seeing animate entities with the minimal features/inputs which trigger the device), our ability to attribute mental states to other things allows us to 'make sense' of much of our world without noticeable effort. It is at this modular level of human cognition that we interpret our world by way of the 'intentional stance' (Dennett 1971, 1987).

It is not, however, at this level of processing that we categorize specific mental states. In addition to detecting others' mental states, we also attribute *particular domains* of knowledge and feelings to other minds (Bering 2002; Bering & Shackelford 2004; Johnson 2005). What makes the human mindreading system particularly remarkable is that it works together with our learned repertoire of different *kinds* of beliefs, desires, and perceptions, as well as our inferences about objects. While there is evidence suggesting that non-human primates have the ability to mentally represent others' mental states, it is uniquely human to be able to know and state the differences between someone's feeling melancholy or sad, for instance, knowing the difference between Beethoven and Copeland, or feeling vindictive and experiencing *schadenfreude* and so forth. Our external environment provides information about our internal environments; we learn how to make distinctions between types of mental states and accredit others with various concerns and understandings. As such, we can attribute very specific kinds of mental states to others that are exclusive to the human experience (e.g., she is performing calculus in her head). Moreover, we attribute particular domains of knowledge to others as well (e.g., he knows a lot about computers). Domains consist of closely related units of information. One intriguing aspect of religion is that people have even less definite evidence of the contents of supernatural agents' minds than we have of each others', yet throughout the world people act in ways which suggest they have confidence in their assumptions about the concerns and wishes of the supernatural agents that inhabit their lives.

5.2.2 Supernatural Minds, Variation, and Counterintuitiveness

In his classic text *The Golden Bough*, Frazer notes the difficulty in attributing belief in God to 'the lower races' (Frazer 2006 [1890], p. 51):

> If we civilised men insist on limiting the name of God to that particular conception of the divine nature which we ourselves have formed, then we must confess that the savage has no god at all. But we shall adhere more closely to the facts of history if we allow most of the higher savages at least to possess a rudimentary notion of certain supernatural beings who may fittingly be called gods, though not in the full sense in which we use the word. That rudimentary notion represents in all probability the germ out of which the civilised peoples have gradually evolved their own high conceptions of deity; and if we could trace the whole course of religious development, we might find that the chain which links our idea of the Godhead with that of the savage is one and unbroken.

It is immediately clear that Frazer's language is ethnocentric, but if anything his sentiments are an attempt to argue, not that 'savage religion' was savage, but that 'high' conceptions of the gods were simply more refined versions of the same concepts found in non-Western societies. He sees an 'unbroken chain' which links the two traditions. This chain — though not by any means unilinear or unidirectional — is the attribution of mental states to agents without readily apparent bodies.

Our religious thinking can be influenced by mundane cognitive operations. Attributing agency to otherwise non-agentive things may be the best bet for an organism's fitness, because failing to detect agency when an agent exists, such as a predator, may mean the organism's demise (Guthrie 1980, 1993, 1995). However, as discussed above, the bulk of our agency attribution is not religious in nature. For example, we readily talk about omnipresent, omniscient gods as though they are not much different than people. Such inconsistencies in thought are considered 'theologically incorrect' (Barrett 1998,1999; Barrett & Keil 1996) insofar as we often process concepts of the gods as though there was nothing particularly supernatural about them at all.

Theologically incorrect thinking is the output of other 'best bet' computations. As such, the distinction between theologically correct and incorrect religious ideas tells us more about the nature of the human mind (and perhaps dogma) than about the nature of religion. We effortlessly attribute agency to material as well as immaterial objects. Statements such as "the University doesn't like it when we drink alcohol on campus" or "the government just wants your money" reflect such a tendency. Our propensity to anthropomorphize has arguably even made it possible for the modern corporation (root: *corpus*) to have the status of a legal person! Not only do we naturally think of collections of people as single agents, but we also design laws enforcing such a trend, allowing the individual constituents of the organization to have limited personal liability for their collective decisions.

What, then, distinguishes 'religious' agency attribution from everyday agency attribution to other bodiless agents such as institutions? One possible answer is that, even though we may readily think in terms of their agency, we can recognize that corporations, universities, and the like are comprised of individuals and lack most features of people, whereas religious concepts are not as easily unpacked as this. In other words, the agency of institutions is a perceived emergent property of a collection of bodies. While there is nothing particularly salient about thinking of a group of people as one person, there is something remarkable about believing in an agent that is not grounded in the empirically verifiable world.

If the tendency to grant minds to so many things is a mundane feature of our species, what then makes concepts religious? Many current approaches characterize religious concepts as essentially 'counterintuitive': these concepts violate deep assumptions we have about our essential categories of objects in the world (Atran 2002; Boyer 1994, 2001; Pyysiäinen 2004). Although these approaches assume that agency is not something we normally attribute to plants or artifacts, we in fact explain what plants and artifacts 'do' in terms of agency all the time. Consider the following statements drawn from Dennett's (1987, p. 59) discussion of the intentional stance:

1. My jade plants appreciate the love I give them.
2. My jade plants prefer Mozart to the Melvins.
3. My jade plants know of all the bad things you did as a child.
4. My jade plants know where the sun is.
5. This block of Wisconsin Cheddar appreciates your fine tastes.
6. The thermostat knows how warm it is in here.
7. "Lightning [...] always wants to find the best way to ground, but sometimes it gets tricked into taking second-best paths" (Dennett 1987, p. 65).

Attributing mental states to a plant because it moves towards sunlight or grows better when it is loved (1) or 'listens' to Mozart or the Melvins (2) is perfectly intuitive when explaining movement and change with a preference and interpreting two sequential events as linked by a causal force (presumably 'listening' to one or the other makes them grow better). Attributing mental states to a stationary block of cheese (5), on the other hand, applies agency to an inanimate entity rendering the statement counterintuitive. Yet the thermostat (6) is inanimate and has effects on the world; it seemingly acts on its own (preprogrammed) accord, as does lightning (7). Indeed, there is an intuitive–counterintuitive continuum (Norenzayan et al. 2006): while (1), (2), and (3) attribute agency to plants and are therefore counterintuitive in the technical sense, (1) and (2) *make intuitive sense*, whereas (3) does not. Suggesting that a plant is cognizant of all the awful things one did as a child does not necessarily violate our basic ontological intuitions; rather it attributes a particular domain of knowledge to plants that they are not normally accredited with.

Compare (3) and (4). In both cases *the capacity for knowledge* is applied to plants, but (4) makes far more intuitive sense than (3). Why? In the case of (3), it is *the particular domain of knowledge attributed* to the jade plant that makes this statement so striking, not the attribution of mental states. It is not, then, the simple attribution of agency to an agentless object that renders an idea counterintuitive (in the technical sense). Since humans are not capable of knowing everything that someone else did wrong, attributing such knowledge to plants is a violation of our expectations about the knowable (i.e., 'counterschematic'; see Barrett 2008; Purzycki 2006, in press a; Purzycki & Sosis 2010). Likewise, if we accredit dogs with the capacity to have such knowledge, it is a violation of our inferences about the knowable, not about animals. Granting a non-agent agency is not necessarily counterintuitive without qualifying what *kind* of agency is attributed to a non-agent and the attributed domains of knowledge. As we discuss below, it is these attributed domains that make supernatural agents particularly salient concepts for people.

The central religious concepts of many traditions are not anthropomorphized supernatural beings (see Guthrie 1980; 1985), but rather supernatural forces or animals. Vine Deloria Jr. (1992) notes that:

> The overwhelming majority of American Indian tribal religions refused to represent deity [sic] anthropomorphically. To be sure, many tribes used the term *grandfather* when praying to God, but there was no effort to use that concept as the basis for a theological doctrine by which a series of complex relationships and related doctrines could be developed. While there was an acknowledgment that the Great Spirit has some resemblance to the role of a grandfather in the tribal society, there was no great demand to have a 'personal relationship' with the Great Spirit in the same manner as popular Christianity has emphasized personal relationships with God (79).

Rather (Deloria Jr. 1979, pp. 152–153; see Powers 1975, pp. 45–47):

> [...] it is with the most common feature of primitive awareness of the world — the feeling or belief that the universe is energized by a *pervading power* [emphasis added]. Scholars have traditionally called the presence of this power *mana*, following Polynesian beliefs, but we find it among tribal peoples, particularly American Indian tribes, as *wakan* [Sioux], *orenda* [Iroquois], or *manitou* [Ojibwe]. Regardless of the technical term, there is general agreement that a substantial number of primitive peoples recognize the existence of a power in the universe that affects and influences them.

Such forces are characterized as creative and intelligent, and often attributed with intentionality in order to transmit ideas more effectively. From a cognitive perspective, however, it is not surprising that Christian missionaries translated such concepts regularly as a personified 'god'. The Sioux concept of *Wakan Tanka*, often translated as 'Sacred Vastness', 'Big Holy', or 'Great Incomprehensibility' (DeMallie 1987, p. 28):

> [...] was the sum of all that was considered mysterious, powerful, or sacred. [...] *Wakan Tanka* never had birth and so could never die. The *Wakan Tanka* created the universe. [...] Rather than a single being, *Wakan Tanka* embodied the totality of existence; not until Christian influences began to affect Lakota belief did *Wakan Tanka* become personified.

However, it seems likely that it was not Christian influence that resulted in American Indians *talking* about sacred forces *as though* they were anthropomorphized and/or attributed with mental states (see Cohen 2007, pp. 104–114 for another example in the case of spirit possession). Evolutionary theorists often do the same when they talk about 'selection', knowing full well that Nature lacks both intentionality and any discriminating taste for the more fit.

In animistic traditions, there are certainly supernatural agents, but not necessarily "culturally postulated *superhuman* agents" (McCauley & Lawson 2002, p. 8; emphasis added). In Tuva, for example, there are many mineral springs (*arzhaannar*) and each spring has its own 'spirit master', as all features of the natural world are believed to be animated by such agents. These spirits take various forms. One of Purzycki's (in press b) informants noted that:

> Everyone prays to the *arzhaannar*. Because they are alive. All of the *arzhaannar* have their spirits. The spirit of Adargan Arzhaan of Sagly [village in southwestern Tuva] is a small marmot. It appears to shamans and lamas. It protects that place. So a man should pray to it. They say there is a bird in this *arzhaan*. It also appears. We notice it in the night when it makes noise. All of these *arzhaannar* have their spirits. That is why every Tuvan prays to his *arzhaannar*, his lands. If we take, for example, Ubsa-Khol [a lake on the border between Tuva and Mongolia], its spirit is a big bull. Each place has its spirit. That is why a Tuvan prays when he is on the road, even if he can't see the spirits. It's the Tuvan people's good ritual.

In this case, spirit masters of these various places are all represented as animal spirits. These spirits animate features of the natural world with life. Even though they cannot see the spirit, people pray to them because these spirits have a protective power over their domains of governance.

To summarize thus far, there is considerable variation in how supernatural agents' forms are represented; some agents are conceived with bodies, others are not, and some supernatural agents are conceived with bodies in some contexts while in other contexts they are believed to be bodiless. It does not appear to be the case that counterintuitiveness or attributions of agency make ideas religious. What makes supernatural agents specifically religious (i.e., worth committing to) are their *attributed* domains of knowledge and concern. In other words, as discussed below, in contrast to other non-agents that we attribute agency to, such as institutions and fictional characters, supernatural agents possess relevant social knowledge and are conceived of as acting upon this knowledge. It is this knowledge and concern which informs — and is informed by — religious behavior.

5.3 Variation in Domains of Supernatural Agents' Knowledge and Concern

5.3.1 Omniscience with Heightened Concern: Prosocial Behavior

> Believing that there are many spirits makes more sense than believing in one god. There are a lot of rivers and mountains. How can one god watch over everything? — Anatoli Kuular, Tuva Republic (Levin 2006, p. 29)

Just as we attribute particular domains of knowledge to other humans, we do the same to our deities. While some supernatural agents know everything we do and think, others are not concerned with such matters, and their knowledge is limited. We can test each others' knowledge about particular domains, but we have no concrete evidence regarding the minds of supernatural agents, let alone what types of knowledge they possess. We find, however, that there are patterns in the mental contents attributed to deities. Some have argued that throughout history people have committed themselves to 'the gods', rather than countless other supernatural beings (e.g., cartoon characters, leprechauns, goblins, etc.) precisely because the gods are accredited with access to valuable social information (Atran 2002; Boyer 2001).

Recent evolutionary theories of religion claim that supernatural agents that evoke religious commitment and devotion are particularly concerned with certain types of social knowledge. This knowledge primarily consists of breaches of prosocial responsibilities (i.e., moral behavior). As such, commitment to supernatural agents may function to inhibit self-interested behavior, and thus in turn contribute to the evolution and persistence of human cooperation (Bering & Johnson 2005; Johnson 2005; Norenzayan & Shariff 2008). Populations differ both in the sets of values they maintain and in the importance they attribute to different types of prosocial behavior, and we would thus expect the concerns of supernatural agents to vary accordingly. What the gods know is an interesting question, but what the gods are *concerned with* is something that is more likely to motivate us to act in socially prescribed ways. The Abrahamic God might not like it if you steal, for instance, but if you live in a small community where little value is held on the accumulation of personal property, then your deity may be more concerned with stinginess.

Boyer (2002, p. 75) argues that it is the perceived access of supernatural agents to socially relevant (i.e., socially strategic) information which makes them salient in our minds. Boyer makes a distinction between agents with 'perfect' and 'imperfect' access to such information. While there is cross-cultural variability, Boyer suggests that supernatural agents are typically granted 'perfect access' to socially strategic information — a very specific domain of all conceivable knowledge. A number of studies have examined the distinctions between what people are supposed to attribute to God, known as theological correctness, and how people actually think about God. Whereas people say that God is omniscient and omnipotent when asked about this explicitly, more subtle measures of how people think about God's powers show that people tend to implicitly attribute certain human limitations to God, such as the inability to be in two places at once (Barrett 1998; Barrett & Keil 1996).

In a response time task, we (Purzycki et al., n.d.) found that individuals took a significantly longer time to respond to questions regarding God's knowledge of *positive*, prosocial behavior than those regarding *negative*, antisocial behavior. Moreover, socially insignificant knowledge (e.g., whether God knows how many pickles there are in Seth's refrigerator?) yielded even longer response times. Despite God's proclaimed omniscience, we seem to process God's knowledge about negative social information more quickly than other knowledge we attribute to God.

Not all supernatural agents, however, are concerned with the general moral behavior of people. For example, when religious traditions are bound to local ecologies, there is a greater stress on sacralizing particular areas which require resource management (e.g., Lansing 2007; Lansing & Kremer 1993) and defense (Sosis, in press). Such agents are acutely concerned with specific behaviors directed towards them in the form of costly rituals. This suggests that there may be no pan-human cognitive bias for supernatural agents concerned about prosocial behavior.

5.3.2 Imperfect Access with Acute Concern: Ritual Behavior

We find significant variation across populations regarding the way people represent their deities' knowledge and concern. Barrett (2002) discusses a number of predictions regarding the relationship between the knowledge and ritual behavior of supernatural agents. If spirits, for example, have imperfect access to human affairs and "can only discern intentions based on a person's actions, then the particular action will have relatively greater importance" than a person's intentions (Barrett 2002, p. 104). On the other hand, "having the right intentions" will be more important during ritual performances directed toward omniscient gods. We suggest that cross-culturally, omniscient supernatural agents will be primarily concerned with general moral behavior, whereas supernatural agents who are limited in their social knowledge of human affairs will be conceived of as acutely concerned with the performance of ritualized acts that are costly to perform. In short, what spirits and gods know may not be nearly as important for religion as what they care about.

For instance, in the highly complex traditional Lakota (Sioux) religion, if one dreamt of the *Wakinyan* (lightning/Thunderbirds/beings) or one of its associates (e.g., rabbits, barn swallows, etc.), one had been chosen by the Thunderbeings to become a *heyoka* — or sacred clown (see Plant 1994; Wallis 1996 for further discussion). Thomas Tyon noted that "the *Wakinyan* often command the man who dreams of them to do certain things" which are typically quite embarrassing for the initiate. If they fail to do whatever they are instructed to by the Thunderbirds, "*Wakinyan* will surely kill them" by lightning strike (Walker 1991, pp. 155–156). In sum, the Thunderbirds will present the dreamer with an embarrassing scenario that he or she must act out in public — in some cases, it is claimed that the conditions and people in the dream are also revealed, making the act quite specific. In this particular case, the supernatural agents — the Thunderbirds — are primarily concerned with whether or not the 'chosen' individual carries out the act as detailed in the dream, and lives as a clown until his or her tenure is completed. Individuals fulfill the wishes of the Thunderbeing to avoid reprisals from them. In this case, specific concentrations of the Sioux supernatural force *Wakan Tanka* (discussed above) are beings accredited with acute concerns and knowledge of the ritual behaviors of those 'chosen' to be clowns.

In Tuva, local 'spirit masters' of specific areas are also not accredited with concern for general human conduct. Rather, they are exclusively concerned with human conduct *towards them*. They are neither concerned with, nor do they punish people for antisocial behavior towards one another, or even for leaving garbage around a sacred site. Although there is no obligation to do so, one pays respects to (i.e., 'feeds') spirit masters by making offerings of food, money, and/or tobacco, as well as by tying a prayer tie to the place where they are honored. Interestingly, there appears to be no consensus regarding the breadth of knowledge of these spirits. Most suggest that spirit masters only know what happens in their areas of governance, and few claim that they are omniscient.

However, there is virtual unanimity when it comes to the question of what spirits care about. After a barrage of questions regarding the moral concerns of spirit masters, one rather exasperated informant told Purzycki (in press b): "They don't *care* about litter, they don't care about how you behave, outside of paying attention to them and 'feeding' them, otherwise they get angry." This suggests that there is not necessarily an evolved, cognitive bias toward representing supernatural agents as morally concerned minds, but rather a necessary flexibility in the domains of knowledge and concern accredited to deities. We also expect these attributed domains of knowledge to correlate with the particular types of behavior prescribed ritually.

In both the Sioux and Tuvan cases, we see the attribution of agency to vague and often inconsistently conceived bodies. A Thunderbeing is often described as "shapeless, but He has wings with four joints each; He has no feet, yet He has huge talons; He has no head, yet has a huge beak with rows of teeth in it" (Walker 1917, cited in Brown 1989 [1953]). The spirit masters in Tuva will frequently manifest themselves in various physical forms, but they are often described as "taking the form of X" rather than being perpetually material. The Abrahamic God is often conceived of as being everywhere, but is attributed a body, not only in present day thinking (Barrett & Keil 1996), but in sacred scriptures as well.

Conceptualizations of these supernatural agents are particular to their respective traditions. Each tradition, however, delimits the range of worldly affairs that these entities are particularly concerned about. Such specific domains of concern are not essential components of our basic ontological categories, and nor can they be produced by innate modules. When we entertain the concept of God, the Thunderbeings, or spirit masters, our mindreading system allows us to attribute a mind to these entities. God concepts and the anthropomorphic spirit masters may violate default expectations about people, and the Thunderbirds and animal spirit masters may violate default expectations about animals. Experimental studies suggest that these violations make such concepts easier to remember than intuitive ideas (Boyer

2000; Boyer & Ramble 2001). However, these supernatural agents vary considerably in their forms, concerns, and abilities. This variance represents differences in our cognitive models or schemas of our particular deities (for further discussion of the distinction between templates and schemas in the context of understanding religious concepts, see Barrett 2008; Purzycki in press a; Purzycki & Sosis 2010). So where and why do we find these divergences between what supernatural agents care about?

5.3.3 Emphases on Faith, Practice, and Social Complexity

Many influential thinkers have claimed that religion results from a deep-rooted need to understand humanity's place in the universe (Darwin 2004/1871; Durkheim 2001/1915, pp. 170–171; Geertz 1973, pp. 108, 140; Russell 1961, pp. 574–575). Elsewhere (Purzycki & Sosis 2009), we suggested that an individual's satisfaction in his or her religious worldviews derives from confirmation by peers. In other words, our religious cohorts confirm our convictions with the predicted prosocial behavior inherent in religious groups. Perceived sharedness has mediating effects on judgment and compliance (Zou et al. in press), and it also affects behavior towards members of out-groups (Sechrist & Stangor 2001, 2007).

However, religious models (i.e., beliefs) may lack consistency between individuals in a religious community. This may be offset by an emphasis on consistency in behavior. We would suspect, then, that under particular conditions, some religious communities will emphasize faith, whereas others will emphasize practice. Fernandez (1965) made a crucial distinction between what he calls *social* and *cultural consensus*. Social consensus is an emphasis on the shared "agreement to orient action towards one another. This acceptance and agreement involves the acceptance of a certain set of signals and signs which give direction and orientation to this interaction permitting the coordination and co-existence of the various participants. A good example of social consensus is found in ritual action" (Fernandez 1965, p. 913). Cultural consensus, on the other hand is an emphasis on shared beliefs or the measurable degree to which individuals share cognitive models (Romney et al. 1986).

In his analysis of a newly emerging trend among the Fang, Fernandez (1965) observed very little cultural consensus regarding the claimed function of their religious rituals. In fact, among his informants he found "a feeling that too great a concern with [cultural consensus] might actually interfere with social consensus — the readiness to orient actions toward one another and engage in ritual activity" (Fernandez 1965, p. 914). It is the sharedness, or at the very least *perceived* sharedness, which needs to be maximized and exploited in order to motivate individuals to cooperate. Nevertheless, without appeals to supernatural agents, such ritual action would not be as long-lived as secular costly rituals (Sosis & Bressler 2003).

On the other hand, outside of a few traditions, faith or belief in religious concepts is often not of fundamental importance. Cohen (2002) found that belief was a strong predictor of life satisfaction, and significantly more so for Christians than for Jews. Also, Cohen et al. (2003) demonstrated that, while Jews and Protestants placed similar emphasis on practice, Protestants were significantly more likely to emphasize faith as an indicator of religiosity. The authors predict that the emphasis on practice and not on faith is probably correlated with how tightly religions are bound to ethnicity (e.g., Hinduism). Faith is a central tenet of Buddhism (see Rahula 1974, p. 8) for example, yet many forms of Buddhism do not endorse the idea of an omniscient deity.

As discussed above, there seem to be two primary domains of behavior that supernatural agents are concerned about: general moral behavior and ritually prescribed behavior. These correlate with emphasis on behavior and belief. These polarities may also correlate with group size. There is an ever-growing literature on the signaling value of religious behaviors (Alcorta & Sosis 2005; Bulbulia 2004; 2009; Henrich 2009; Sosis 2005; Sosis & Alcorta 2003; Sosis & Bressler 2003). The results of such studies suggest that religion evolved to overcome the inherent challenges of cooperation. Human communities are vulnerable to individuals who may exploit others for personal gain. If some members of a group shirk their duties, yet reap the benefits of others' work, the community may ultimately become overrun by exploiters. Religious traditions provide the rationalizations and motivations to engage in the ritualized behaviors that can signal commitment, thereby increasing trust and cooperation within communities. In traditional societies, religion was not something experienced on a particular day or during particular times; it was integrated with all domains of human experience. Large societies complicate these patterns.

Stark (2001) demonstrates that moralizing gods are found primarily among large societies with higher degrees of economic specialization (i.e., agricultural). The more complex a society is, the more likely a population is to worship a high, moralizing deity (Johnson 2005; Lahti 2009; Rappaport 1999; Sanderson 2008; Swanson 1960; Wallace 1966). As group size increases and occupations become more specialized, religion also becomes more diversified, institutional, compartmentalized, and doctrinal (Boyer 2001; Whitehouse 2004). While there is variation in religious thought and practice in non-state societies, there are fewer competing traditions than in state-level societies. As the size of a population grows, social accountability is impaired, and thus the form of a population's religion must change to counter the problems of religious diversity, anonymity, and accountability. It becomes more taxing for communities in larger populations to monitor commitment. Someone may reap the benefits of the group, and when threatened with sanctions for not reciprocating, he or she may simply seek opportunities elsewhere. Yet badges of religious affiliation are reliable signals of trustworthiness to individuals, even though members might not know each other, or may even be of the same tradition (Sosis 2005). Omniscient deity concepts are attempts to curb such problems of social complexity. Emphasis on cultural consensus at the expense of social consensus probably emerges in contexts where religious traditions are:

- intertwined with imperial expansion,
- not an ethnic religion,
- not rooted in a specific ecological area, and/or
- in competition with one or more traditions.

5.4 Conclusion

This chapter began with a description of the pan-human ability to attribute mental states to all sorts of entities. There is abundant variation in the *kinds* of supernatural agents we believe in, but limited variation in the types of *knowledge* we attribute to them. The knowledge and concern attributed to supernatural agents seem to vary across two domains: moral actions and ritual actions. Any evolutionary account of religion must be able to explain the considerable cross-cultural variation in religious expression and belief. While the religious system can be maladaptive (e.g., suicide cults, exclusive reliance on faith healing, etc.), if it evolved to promote cooperation, then variation will only be sustained inasmuch as it effectively overcomes problems of defection. The ability to monitor the behavior of other group members will influence the concerns of a community's supernatural agents. Nonetheless, supernatural agents concerned with either moral or ritual actions are likely to motivate cooperative behavior among constituents in the form of social support. The concerns of supernatural agents, whether moral, ritual, or both, will be systematically associated with other elements that comprise the religious system, including the importance of faith, practice, ethnicity, proselytization, and ecology. Given particular constraints, the religious system will respond to diverse socio-ecological conditions and generally adapt to ensure group cohesion and prosocial behavior.

Acknowledgements We thank Oxford University's Cognition, Religion, and Theology Project, the John Templeton Foundation, and the University of Connecticut for generous support. We also thank Jordan Kiper, John Shaver, and Paul Swartwout for their critical readings of earlier drafts of this chapter. Purzycki would like to thank Jessica McCutcheon, Anai-Xaak and Zhenya Saryglar, Valentina Süzükei, and all his Tuvan informants.

References

Alcorta CS, Sosis R (2005) Ritual, emotion, and sacred symbols. Human Nature 16(4):323–359
Atran S (2002) *In Gods We Trust: The Evolutionary Landscape of Religion*. Oxford University Press, Oxford
Atran S, Norenzayan A (2004) Religion's evolutionary landscape: Counterintuition, commitment, compassion, communion. Behavioral and Brain Sciences 27(6):713–770
Baron-Cohen S (1997) *Mindblindness: An Essay on Autism and Theory of Mind*. The MIT Press, Cambridge
Barrett JL (1998) Cognitive constraints on Hindu concepts of the divine. Journal for the Scientific Study of Religion 37:608–619
Barrett JL (1999) Theological correctness: Cognitive constraint and the study of religion. Method and Theory in the Study of Religion 11:325–339
Barrett JL (2002) Dumb gods, petitionary prayer, and the cognitive science of religion. In Pyysiänen I, Veikko A (eds) *Current Approaches in the Cognitive Science of Religion*. Continuum, New York
Barrett JL (2004) *Why Would Anyone Believe in God?* AltaMira Press, New York
Barrett JL (2008) Coding and quantifying counterintuitiveness in religious concepts: Theoretical and methodological reflections. Method & Theory in the Study of Religion 20:308–338

Barrett JL, Keil FC (1996) Conceptualizing a nonnatural entity: Anthropomorphism in god concepts. Cognitive Psychology 31:219–247

Bering JM (2002) The existential theory of mind. Review of General Psychology 6(1):3–24

Bering JM, Johnson DDP (2005) "O Lord ... you perceive my thoughts from afar": Recursiveness and the evolution of supernatural agency. Journal of Cognition and Culture 5:118–142

Bering JM, McLeod KA, Shackelford TK (2005) Reasoning about dead agents reveals possible adaptive trends. Human Nature 16:360–381

Boyer P (1994) *The Naturalness of Religious Ideas: A Cognitive Theory of Religion.* University of California Press, Berkeley

Boyer P (1996) What makes anthropomorphism natural: Intuitive ontology and cultural representations. Journal of the Royal Anthropological Institute 2(1):83–97

Boyer P (2000) Functional origins of religious concepts: Ontological and strategic selection in evolved minds. Journal of the Royal Anthropological Institute 6(2):195–214

Boyer P (2001) *Religion Explained: The Evolutionary Origins of Religious Thought.* Basic Books, New York

Boyer P (2002) Why do gods and spirits matter at all? In Pyysiänen I, Veikko A (eds) *Current Approaches in the Cognitive Science of Religion.* Continuum, New York

Boyer P, Bergstrom B (2008) Evolutionary perspectives on religion. Annual Review of Anthropology 37:111–130

Boyer P, Ramble C (2001) Cognitive templates for religious concepts: Cross-cultural evidence for recall of counterintuitive representations. Cognitive Science 25:535–564

Brown JE (1989 [1953]) *The Sacred Pipe: Black Elk's Account of the Seven Rites of the Oglala Sioux.* MJF Books, New York

Bulbulia J (2009) Why 'costly-signalling' models of religion require cognitive psychology. In Geertz A, Jensen J (eds) *Origins of Religion, Cognition, and Culture.* Equinox, London

Bulbulia J (2008) Meme infection or religious niche construction? An adaptationist alternative to the cultural maladaptationist hypothesis. Method and Theory in the Study of Religion 20:67–107

Bulbulia J (2004) Religious costs as adaptations that signal altruistic intention. Evolution and Cognition 10(1):19–38

Bulbulia J, Frean M (2009) Religion as superorganism: On David Sloan Wilson's Darwin's Cathedral (2002) in Stausberg, M (ed) *Contemporary Theories of Religion: A Critical Companion.* Routledge, New York

Bulbulia J, Mahoney A (2008) Religious solidarity: The hand-grenade experiment. Journal of Cognition and Culture 8(3-4):295–320

Call J, Tomasello M (2008) Does the chimpanzee have a theory of mind? 30 years later. Trends in Cognitive Sciences 12(5):187–192

Cohen AB (2002) The importance of spirituality in well-being for Jews and Christians. Journal of Happiness Studies 3:287–310

Cohen AB, Siegel JI, Rozin P (2003) Faith versus practice: Different bases for religiosity judgments by Jews and Protestants. European Journal of Social Psychology 33(2):287–295

Cohen E (2007) *The Mind Possessed: The Cognition of Spirit Possession in an Afro-Brazilian Religious Tradition.* Oxford University Press, New York

Darwin C (2004 [1871]) *The Descent of Man.* Penguin Classics, New York

Deloria Jr. V (1979) *The Metaphysics of Modern Existence.* Harper & Row Publishers, New York

Deloria Jr. V (1992) *God is Red: A Native View of Religion.* Fulcrum Publishing, Golden

DeMallie RJ (1987) Lakota belief and ritual in the nineteenth century. In DeMallie, RJ, Parks DR (eds) *Sioux Indian Religion.* University of Oklahoma Press, Norman

Dennett DC (1971): Intentional systems. The Journal of Philosophy 68(4):87–106

Dennett DC (1987) *The Intentional Stance.* The MIT Press, Cambridge

Durkheim E (2001 [1915]): *The Elementary Forms of Religious Life.* Oxford University Press, New York

Feierman JR (2009) *The Biology of Religious Behavior: The Evolutionary Origins of Faith and Religion.* Praeger, Santa Barbara

Fernandez JW (1965) Symbolic consensus in a Fang reformative cult. American Anthropologist 67(4):902–929

Fodor J (1983) *The Modularity of Mind: An Essay on Faculty Psychology*. MIT Press, Cambridge

Fodor J (2000) *The Mind Doesn't Work that Way: The Scope and Limits of Computational Psychology*. MIT Press, Cambridge

Frazer JG (2006) *The Golden Bough: A Study of Magic and Religion*. NuVision Publications, Sioux Falls

Geertz C (1973) *The Interpretation of Culture*. Basic Books, New York

Gelman R, Durgin F, Kaufman L (1995): Distinguishing between animates and inanimates: Not by motion alone. In Sperber D, Premack D, Premack A (eds) *Causal Cognition: A Multidisciplinary Debate*. Plenum Press, Oxford

Guthrie SE (1980) A cognitive theory of religion. Current Anthropology 21(2):181

Guthrie SE (1995) *Faces in the Clouds: A New Theory of Religion*. Oxford University Press, Oxford

Haley KJ, Fessler DMT (2005) Nobody's watching? Subtle cues affect generosity in an anonymous economic game. Evolution and Human Behavior 26(3):245–256

Henrich J (2009): The evolution of costly displays, cooperation, and religion: Credibility enhancing displays and their implications for cultural evolution. Evolution and Human Behavior 30(4):244–260

Johnson DDP (2005): God's punishment and public goods: A test of the supernatural punishment hypothesis in 186 world cultures. Human Nature 16:410–446

Johnson DDP, Bering JM (2006): Hand of God, mind of man: Punishment and cognition in the evolution of cooperation. Evolutionary Psychology 4:219–233

Kirkpatrick LA (2006) Religion is not an adaptation. In McNamara, P (ed) *Where Man and God Meet: How Brain and Evolutionary Studies Alter Our Understanding of Religion*. Vol. 1: *Evolution, Genes, and the Religious Brain*. Praeger, Westport

Kirkpatrick LA (2008) Religion is not an adaptation: Some fundamental issues and arguments. In Bulbulia J, Sosis R, Harris E, Genet C, Genet R, Wyman, K (eds) *The Evolution of Religion: Studies, Theories, and Critiques*. Collins Foundation Press, Santa Margarita

Lahti DC (2009) The correlated history of social organization, morality, and religion. In Voland E, Schiefenhövel W (eds) *The Evolution of Religious Mind and Behavior*. Springer, New York

Lansing JS (2007) *Priests and Programmers: Technologies of Power in the Engineered Landscape of Bali*. Princeton University Press, Princeton

Lansing JS, Kremer JN (1993) Emergent properties of Balinese water temple networks: Coadaptation on a rugged fitness landscape. American Anthropologist 95(1):97–114

Levin T (with Süzükei V) (2006) *Where Rivers and Mountains Sing: Sound, Music, and Nomadism in Tuva and Beyond*. Indiana University Press, Bloomington

McCauley RN, Lawson ET (2002) *Bringing Ritual to Mind: Psychological Foundations of Cultural Forms*. Cambridge University Press, Cambridge

Norenzayan A, Shariff AF (2008) The origin and evolution of religious prosociality. Science 322:58–62

Norenzayan A, Atran S, Faulkner J, Schaller M (2006) Memory and mystery: The cultural selection of minimally counterintuitive narratives. Cognitive Science 30:1–30

Plant J (1994) *Heyoka: Die Contraries und Clowns der Plainsindianer*. Berlag für Amerikanistik, Germany

Powers WK (1975) *Oglala Religion*. University of Nebraska Press, Lincoln

Premack DG, Woodruff G (1978) Does the chimpanzee have a theory of mind? Behavioral and Brain Sciences 1:515–526

Purzycki BG (2006) *Myth, Humor, and Ontological Templates: A Study of the Retention and Transmission of Mythical Religious Ideas*. Unpublished MA thesis, University of Nebraska, Lincoln

Purzycki BG (in press a) Humor as violations and deprecation: A cognitive anthropological account. Journal of Cognition and Culture 10(1/2)

Purzycki BG (in press b) Spirit masters, ritual cairns, and the adaptive religious system in Tyva. Sibirica

Purzycki BG, Finkel DN, Shaver J, Wales N, Cohen AB, Sosis R (submitted) What does God know? Supernatural agents' perceived access to socially strategic and nonstrategic information

Purzycki BG, Sosis R (2009) The religious system as adaptive: Cognitive flexibility, public displays, and acceptance. In Voland E, Schiefenhövel W (eds) *The Biological Evolution of Religious Mind and Behavior*. Springer, New York

Purzycki BG, Sosis R (in press) Religious concepts as necessary components of the adaptive religious system. In Frey U (ed) *Evolution and Religion: The Natural Selection of God*. Tectum Verlag, Marburg

Pyysiäinen I (2004) *Magic, Miracles, and Religion: A Scientist's Perspective*. AltaMira Press, Walnut Creek

Rappaport RA (1999) *Ritual and Religion in the Making of Humanity*. Cambridge University Press, Cambridge

Rahula W (1974) *What the Buddha Taught*, revised edn. Grove Press, New York

Romney AK, Weller SC, Batchelder WH (1986) Culture as consensus: A theory of cultural and informant accuracy. American Anthropologist 88(2):313–338

Russell B (1961) *The Basic Writings of Bertrand Russell*. Simon & Schuster, New York

Sanderson SK (2008) Religious attachment theory and the biosocial evolution of the major world religions. In Bulbulia J, Sosis R, Harris E, Genet R, Genet C, Wyman K (eds) *The Evolution of Religion: Studies, Theories, and Critiques*. Collins Foundation Press, Santa Margarita

Sechrist GB, Stangor C (2001) Perceived consensus influences intergroup behavior and stereotype accessibility. Journal of Personality and Social Psychology 80(4):645–654

Sechrist GB, Stangor C (2007) When are intergroup attitudes based on perceived consensus information? The role of group familiarity. Social Influence 2(3):211–235

Shariff AF, Norenzayan A (2007) God is watching you: Priming God concepts increases prosocial behavior in an anonymous economic game. Psychological Science 18:803–80

Smetana JG (2006) Social–cognitive domain theory: Consistencies and variations in children's moral and social judgments. In Killen M, Smetana J (eds) *Handbook of Moral Development*. Lawrence Erlbaum Associates, Publishers, Mahwah

Sosis R (2005) Does religion promote trust? The role of signaling, reputation, and punishment. Interdisciplinary Journal of Research on Religion 1(7):1–30

Sosis R (2006) Religious behaviors, badges, and bans: Signaling theory and the evolution of religion. In McNamara P (ed) *Where God and Science Meet: How Brain and Evolutionary Studies Alter Our Understanding of Religion*. Vol. I: *Evolution, Genes, and the Religious Brain*. Prager Publishers, Westport

Sosis R (2009a) The adaptationist-byproduct debate on the evolution of religion: Five misunderstandings of the adaptationist program. Journal of Cognition and Culture 9:339–356

Sosis R (2009b) Why are synagogue services so long? An evolutionary examination of Jewish ritual signals. In Golberg R (ed) *Judaism in Biological Perspective: Biblical Lore and Judaic Practices*. Paradigm Publishers, Boulder

Sosis R (submitted) Why sacred lands are not indivisible: The cognitive foundations of sacralizing land

Sosis R, Alcorta C (2003) Signaling, solidarity, and the sacred: The evolution of religious behavior. Evolutionary Anthropology 12:264–274

Sosis R, Bressler ER (2003) Cooperation and commune longevity: A test of the costly signaling theory of religion. Cross-Cultural Research 37(2):211–239

Sperber D (1996) *Explaining Culture: A Naturalistic Approach*. Wiley Blackwell, Malden

Stark R (2001) Gods, rituals, and the moral order. Journal for the Scientific Study of Religion 40(4):619–636

Swanson GE (1960) *The Birth of the Gods: The Origin of Primitive Beliefs*. University of Michigan Press, Ann Arbor

Walker JR (1917) The sun dance and other ceremonies of the Oglala division of the Teton Dakota. Anthropological Papers of the American Museum of Natural History XVI, Part II, New York

Walker JR (1991) Lakota belief and ritual. DeMallie RJ, Jahner EA (eds). University of Nebraska Press, Lincoln

Wallace AFC (1966) *Religion: An Anthropological View*. McGraw-Hill, New York
Wallis WD (1996) *Heyoka: Lakota Rites of Reversal*. Lakota Books, Kendall Park
Whitehouse H (2004) *Modes of Religiosity*. Alta Mira Press, New York
Wilson DS (2002) *Darwin's Cathedral: Evolution, Religion, and the Nature of Society*. University
    Of Chicago Press, Chicago
Wilson EO (1978) *On Human Nature*. Bantam Books, New York
Zou X, Tam K, Morris MW, Lee, S, Lau IY, Chiu, C (in press) Culture as common sense: Percei-
    ved consensus vs. personal beliefs as mechanisms of cultural influence. Journal of Personality
    and Social Psychology. SSRN preprint. http://ssrn.com/abstract=1392170. Ac-
    cessed 15 September 2009

Chapter 6
Our Preferences: Why We Like What We Like

Karl Grammer and Elisabeth Oberzaucher

Abstract Humans tend to judge and sort their social and non-social environment permanently into a few basic categories: 'likes' and 'don't likes'. Indeed we have developed general preferences for our social and non-social environment. These preferences can be subsumed under the term 'evolutionary aesthetics' (Voland & Grammer 2003). Indeed humans and animals have evolved preferences for mates, food, habitats, odors, and objects. Those stimuli that promoted reproductive success are bound to evoke positive emotional responses, and humans develop an obsession-like attitude towards aesthetics and beauty. Although we are 'all legally equal', people are often treated differently according to their physical appearance. This differential treatment by others starts early in life. Three-month-old children gaze longer at attractive faces than at unattractive faces. From these results, Langlois et al. (1990) conclude that beauty standards are not learned, but that there is an innate beauty detector. Attractive children receive less punishment than unattractive children for the same types of misbehavior. Differential treatment goes on at school, college, and even university (Baugh & Parry 1991). In this part of our lives attractiveness is coupled to academic achievements — attractive students receive better grades. Even when we apply for jobs, appearance may dominate qualification (Collins & Zebrowitz 1995). This differential treatment reaches its peak perhaps in jurisdiction, where attractiveness can lead to better treatment and lighter sentences. However, this is only the case if attractiveness did not play a role in the crime (Hateld & Sprecher 1986). We even believe that attractive people are better — 'what is beautiful is good' is a common standard in our thinking, according to Dion et al. (1972). The question then arises: Where does this obsessive preoccupation with beauty and attractiveness come from? We will outline here the thesis that human mate selection

Karl Grammer
Department of Anthropology, University of Vienna, Althanstrasse 14, 1090 Vienna, Austria, e-mail: urban.ethology@univie.ac.at

Elisabeth Oberzaucher
Department of Anthropology, University of Vienna, Althanstrasse 14, 1090 Vienna, Austria, e-mail: elisabeth.oberzaucher@univie.ac.at

U.J. Frey et al. (eds.), *Essential Building Blocks of Human Nature*, The Frontiers Collection, DOI 10.1007/978-3-642-13968-0_6, © Springer-Verlag Berlin Heidelberg 2011

criteria, which have evolved through human evolutionary history, are responsible for shaping our perception of attractiveness and beauty.

6.1 Darwin's Problem

Darwin himself already promoted this idea and laid the foundation for any theoretical development in this direction. With regard to a general sense of beauty, Darwin writes in the *Descent of Man* (p. 99):

> This sense has been declared to be peculiar to man. I refer here only to the pleasure given by certain colours, forms, and sounds, and which may fairly be called a sense of the beautiful; with cultivated men such sensations are, however, intimately associated with complex ideas and trains of thought.

And:

> Why certain bright colours should excite pleasure cannot, I presume, be explained, any more than why certain flavours and scents are agreeable; but habit has something to do with the result, for that which is at first unpleasant to our senses, ultimately becomes pleasant, and habits are inherited. With respect to sounds, Helmholtz has explained to a certain extent on physiological principles, why harmonies and certain cadences are agreeable. But besides this, sounds frequently recurring at irregular intervals are highly disagreeable, as everyone will admit who has listened at night to the irregular flapping of a rope on board ship.

And further on:

> The same principle seems to come into play with vision, as the eye prefers symmetry or figures with some regular recurrence. Patterns of this kind are employed by even the lowest savages as ornaments; and they have been developed through sexual selection for the adornment of some male animals. [...] Whether we can or not give any reason for the pleasure thus derived from vision and hearing, yet man and many of the lower animals are alike pleased by the same colours, graceful shading and forms, and the same sounds.

Darwin already mentions all the principles used as explanatory principles in aesthetic perception today. Although Darwin grasped the basic principle, he had no explication for the sense of beauty in animals and man. When reading the *Descent of Man*, the reader finds Darwin wandering between the assumptions that evolution must have worked on aesthetic preferences in both animals and humans, and a general cultural variation of these aesthetic preferences. After reviewing the evidence he had at this time, he came to the conclusion (p. 581):

> It is certainly not true that there is in the mind of man any universal standard of beauty with respect to the human body. It is, however, possible that certain tastes may in the course of time become inherited, though there is no evidence in favour of this belief: and if so, each race would possess its own innate ideal standard of beauty. It has been argued that ugliness consists in an approach to the structure of the lower animals, and no doubt this is partly true with the more civilised nations, in which intellect is highly appreciated; but this explanation will hardly apply to all forms of ugliness. The men of each race prefer what they are accustomed to; they cannot endure any great change; but they like variety, and admire each characteristic carried to a moderate extreme. Men accustomed to a nearly oval

face, to straight and regular features, and to bright colours, admire, as we Europeans know, these points when strongly developed. On the other hand, men accustomed to a broad face, with high cheek-bones, a depressed nose, and a black skin, admire these peculiarities when strongly marked. No doubt characters of all kinds may be too much developed for beauty. Hence a perfect beauty, which implies many characters modied in a particular manner, will be in every race a prodigy. As the great anatomist Bichat long ago said, if every one were cast in the same mould, there would be no such thing as beauty. If all our women were to become as beautiful as the Venus de Medici, we should for a time be charmed; but we should soon wish for variety; and as soon as we had obtained variety, we should wish to see certain characters a little exaggerated beyond the then existing common standard.

Today, within cultures the generality of attractiveness is easily accepted. Several rating studies, especially those by Iliffe (1960) have shown that people of an ethnic group share common attractiveness standards. In this standard, beauty and sexual attractiveness seem to be the same, and ratings of pictures show a high congruence over social class, age, and sex. Thus it seems to be a valid starting point when we state that beauty standards are at least shared within a population. Moreover, more recent studies (Cunningham et al. 1995) suggest that the constituents of beauty are neither arbitrary nor culture bound. The consensus on which women are considered to be good-looking and which are not is quite high between four cultures (Asian, Hispanic, black, and white women rated all by males from all cultures). Thus Darwin was not completely right in his argumentation on the possible uniformity of beauty perception with respect to the human body.

Nevertheless, Darwin argues that aesthetic principles in human perception result from sexual selection (p. 583):

> We must next inquire whether this preference and the consequent selection during many generations of those women, which appear to the men of each race the most attractive, has altered the character either of the females alone, or of both sexes. With mammals the general rule appears to be that characters of all kinds are inherited equally by the males and females; we might therefore expect that with mankind any characters gained by the females or by the males through sexual selection would commonly be transferred to the offspring of both sexes.

Why did Charles Darwin come to this somewhat fuzzy conclusion? With the information available then, the conclusion was surely right. He had asked missionaries and ethnographers to describe the beauty standards of different ethnic groups, resulting in a diversity of answers that made generalization difficult. Even today, we might accept Darwin's conclusions considering the wide diversity of human appearance. And more than 100 years later, it still seems obvious for some researchers that beauty standards are culturally determined. As an example, Grogan (1999) points out that beauty standards vary greatly over time, through history and between (and even within) societies. This high degree of cultural relativism culminates in the following statement:

> Evolutionary psychologists have failed to demonstrate convincingly that preferences for particular body shapes are biologically based [...]. Current data suggest that body satisfaction is largely determined by social factors, and is intimately tied to sexuality.

In this statement Grogan (1999) neglects the ultimate connection between sexuality and reproduction.

This line of argumentation has produced almost no scientific progress since the time of Darwin. If human sexuality is linked to human reproduction, evolutionarily determined constraints should be highly effective. Although we have to admit that beauty standards might differ between cultures and times, we will show in this chapter that the underlying constraints which shaped the standards remain the same, and that these constraints may be of an evolutionary origin.

Darwin's inability to explain the foundations of aesthetic preferences has been resolved by modern evolutionary theories. It was not until the 1960s that the general theoretical foundation for the explanation of our aesthetic preferences finally emerged.

It was basically Konrad Lorenz who laid the foundation for a deeper understanding of the possible mechanisms that could be responsible for our aesthetic preferences. The first idea is connected to the fact that the cognitive apparatus of humans evolved to fit the actually present physical structures, and that our thinking and reasoning are a result of evolution. This means that the cognitive structures we have are adaptations to problem-solving in the past. The brains — or cognitive algorithms — that were able to process the information from our environment more efficiently were those that were selected in the course of evolution. But Lorenz went beyond that. He argued that, in the course of evolution, certain environmental stimuli which promoted reproductive success were connected to positive emotional reactions (Lorenz, 1973).

This idea has recently received more attention from a new discipline called evolutionary psychology. Extending Lorenz' adapted mind theory, Cosmides and Tooby (1992) propose that our brains have domain-specific cognitive structures which evolved in order to process everyday information efficiently and check the impact of this information on reproductive success. Is there any evidence that we have such domain-specific adaptations in reasoning regarding aesthetic principles? When Aharon et al. (2001) showed photographs of attractive and unattractive people to males, they could demonstrate that, in all males, the same brain regions were active when viewing attractive faces.

6.2 Evolutionary Constraints on Aesthetic Perception: The Body as an Evolved Form

If adaptations to the perception of beauty and for aesthetic preferences exist as modules of the mind (Cosmides & Tooby 1992), they should follow evolutionary constraints that affect reproductive success. Host–parasite co-evolutionary cycling (Hamilton & Zuk 1982) predicts that parasite resistance should be a valued trait in mate selection. One defense against parasites is the production of high degrees of polymorphism: when a parasite adapts to one genetic allele, alternative alleles may be advantageous. Pathogens are the major environmental perturbations causing developmental instability, and developmental stability may be related to high genetic variance in disease resistance, which in turn may relate to fitness. In this case

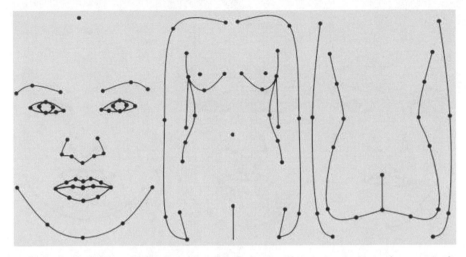

Fig. 6.1 Averages of landmarks from female faces and bodies calculated with modern geometric morphometrical methods. Averages are usually more beautiful than the single parts constituting them. From Schaefer et al. (2006)

one problem arises: the possibility of deception. Zahavi and Zahavi (1997) propose that the honesty of signals might be triggered by a handicap. This means that the generation of the signal itself is costly and imposes a handicap on the signal bearer.

For instance the attractive broad chin in males could signal an immune handicap. This is the only male facial feature where a positive correlation with attractiveness has been replicated several times: 'wide jaws and big chins', and generally bigger, lower faces (Mueller & Mazur 1997; Grammer & Thornhill 1994). This feature might be attractive because high testosterone levels are required to produce it. The costs of high testosterone levels might lie in the production costs themselves, or might be the testosterone-induced suppression of the immune function with resulting increased disease susceptibility during puberty (Folstad & Karter 1992). Immunocompetence is highly relevant because the steroid reproductive hormones may impact immune function in a negative way (Folstad & Karter 1992). Extreme male features, which are triggered by testosterone, thus advertise honestly that their bearer's immune system was sufficiently parasite-resistant to outweigh the handicap. Yet women modify their perception of male traits according to their menstrual cycle. At the point of highest conception probability (ovulation), the typical male sex-hormone markers are most attractive (Johnston et al 2001). Sex hormone markers also play a role for female beauty. The typical female fat distribution (i.e., breasts, buttocks, and lip height) manifests itself under the effect of estrogen. But it is not only the size of sex hormone markers, but also their construction which counts for attractiveness ratings. For instance, female breasts with a V-shaped breast axis are perceived as most attractive by men (Grammer et al. 2001). Female fat distribution may also signal a stable hormonal state, because fat cells can store estrogen and thus stabilize the female cycle.

The general blueprint for the development of sex hormone markers is developed very early in life. During the first six weeks of gestation, the mother's sex hormones,

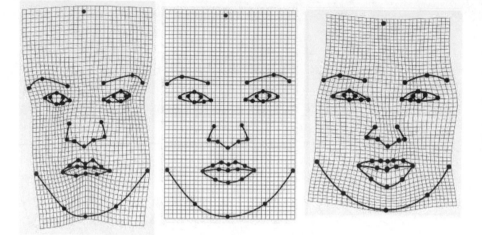

Fig. 6.2 Attractive faces (*right*) show less fluctuating asymmetry than unattractive faces (*left*). The *center picture* is the average. From Schaefer et al. (2006)

both estrogen and testosterone, are responsible for later masculine or feminine facial appearance (Fink et al. 2005).

6.3 The Eight Pillars of Beauty

In general human attractiveness seems to signal genetic fitness and health (Grammer et al. 2005). Besides sex hormone markers, we find additional elements of a construction set for beauty templates. Youthfulness seems to be paramount and signal reproductive potential (Grammer et al. 2003a). One sign for youth is the absence of body hair or the presence of child-like features — both traits contribute to high attractiveness ratings. Men also show a preference for blonde hair. This also seems to be a sign of youth, since throughout ontogeny human hair darkens with age (Grammer et al. 2001).

Besides youth, we find that fluctuating asymmetry, i.e., randomly occurring asymmetries in the body and face, is unattractive (Grammer & Thornhill 1994). The absence of asymmetries could signal developmental stability. This means that missing asymmetry might be a sign that the organism was able to deal with environmental perturbations during development. This phenomenon can also be found in voices. Attractive voices are linked to low asymmetries in bodies (Hughes et al. 2002). Thus symmetry may be an honest signal for mate quality. Symmetry has been shown to be a mate selection criterion in many species, from scorpion flies (Thornhill 1992) through birds (Møller 1992) to humans.

The relation between attractiveness and most body measurements is not linear, and most of the traits show highest attractiveness at average sizes. Generally, ave-

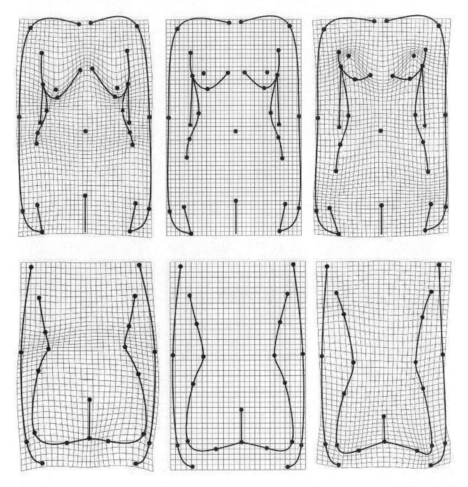

Fig. 6.3 Attractive front and back views (*right*) show less fluctuating asymmetry and determined sex hormone markers (*left*). The *center picture* is the average. From Schaefer et al. (2006)

rages seem to be more beautiful, not only in faces, but in most body measures. Humans tend to avoid extremes and this also accounts for attractiveness (Grammer & Thornhill 1994). For instance, humans prefer average breast sizes and average waist to hip ratios in women.

Another pillar of beauty is body odor. Humans have a genetically determined individual signature in their body odor (Penn et al. 2006) that conveys information about the immune system. Presuming that parasite resistance is a crucial trait, the immune system could advertise its capacity via body odor. Indeed symmetrical women with attractive faces also smell pleasant for men (Rikowski & Grammer 1999). Women modulate their preferences in the course of the menstrual cycle — attractive, symmetrical men smell pleasant only at mid-cycle.

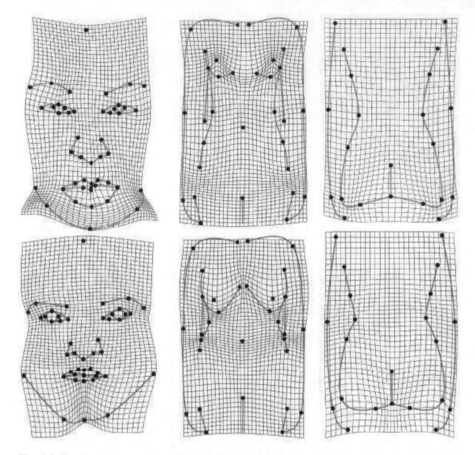

Fig. 6.4 Simultaneous regression of the whole body in different views reveals that the whole body is one single ornament (attractive *upper row*). From Schaefer et al. (2006)

An almost completely neglected pillar of beauty is the appearance of skin texture. With the help of computer-based image analysis of skin texture, we showed that homogeneous textures are more attractive and signal youth (Fink et al. 2006; Mas et al. 2007). Furthermore, skin texture homogeneity and attractiveness are linked to a variable, parasite-resistant immune system (Roberts et al. 2005).

Empirical evidence for the relevance of beautiful hair is rather scarce. We know that long and shiny hair is attractive (Grammer et al. 2001), and we can speculate that long hair is extremely difficult to produce, so its shininess could signal parasite resistance and thus genetic fitness.

The last pillar is body motion. Evidence shows that body motion is attractive when it is sex specific (Grammer et al. 2003b), and attractive dancers are highly symmetrical (Brown et al. 2005).

This brings us to a general problem: How do the traits interact? We assume that the human body is one ornament and that the attractiveness of traits is correlated.

Indeed, women with beautiful faces also have beautiful bodies (Thornhill & Grammer 1999). And from the results presented above, we also know that symmetry is related to most features, like body odor, voice attractiveness, and facial or bodily attractiveness. Thus we can conclude that the pillars of beauty are interrelated and form an ornament. This ornament signals genetic fitness.

But there are several caveats for an approach like this: 'attractiveness' has to be a flexible concept. The reason for this is that a fixed template for attractiveness could unnecessarily narrow down the possibilities in mate selection. This brings up another question: Why do we not become more and more attractive and beautiful through sexual selection? This notion gives rise to an argument already presented here. Van Valen (1973) called the argument the red queen hypothesis. It is based on one of Alice's experiences in Lewis Carroll's novel *Alice Through the Looking Glass*. Alice proposes to race the red queen across the chessboard in order to become queen herself. Unfortunately, one of the principles in the world through the looking glass is that you have to run twice as fast as you can simply in order to move and leave your place. Host–parasite co-evolution can be considered as just such a race. In this view, the only answer lies in genetic variability which can promote a certain time deferment in the race against an unpredictably changing environment. Thus we are able to go well beyond Darwin in our argumentation — attractiveness is a set of construction templates that can be assessed empirically.

6.4 Alternative Views: Neuroaesthetics

Several studies have repeatedly shown (Galton 1879; Kalkofen et al. 1990; Langlois & Roggman 1990; Müller 1993, Grammer & Thornhill 1994; Perrett et al. 1994) that computer-generated prototypical faces are more attractive than the single faces used to generate them. But there are two caveats: this is only true for female faces and all researchers find that there are some individual faces that are more attractive than the prototypes.

If our brain uses prototypes, the aesthetic preference for averageness in human faces could well be coupled to being 'prototypical'. Thus there might be a better fit of the stimulus onto the prototypical face template underlying cognitive processing. As a result, prototypes are recognized faster and better and thus might create higher arousal. This fact could be the reason for the preference for averageness. Our brain may more willingly accept better-fitting stimuli. Müller (1993) has called this process neuroaesthetics. This approach is carried further by Enquist and Arak (1994). These authors propose that any notion of beauty or aesthetics is not linked to biological function in the sense that such preferences are related to reproductive success. In contrast they see aesthetic preferences as a result of the general construction of the perceptive and cognitive apparatus that has to follow certain construction rules.

In recent years, fMRI studies have been conducted in order to identify the brain structures that are assumed to control facial perception and the processing of attractiveness. Ishai (2006) and Kranz and Ishai (2006) scanned subjects during a facial attractiveness assessment task. These studies showed that the processes involve a network of face-specific visual, limbic, and prefrontal regions. The second study indicates that the way faces are perceived is modulated by sexual preference. In particular, the orbitofrontal cortex seems to be especially sensitive to happy expressions and attractive faces (O'Doherty et al. 2003). Ishai (2006) showed that these structures are active independently of the sex of the stimulus faces and the sexual orientation of the subject. On this basis the author concluded that the processing of attractive faces is independent of reproductive value, because there is no dissociation between males seeing females and vice versa, and she concludes that sexual preference modulates neural responses to relevant stimuli in the adult brain, rather than reproductive fitness. The study itself actually provides no evidence for such a statement, because this was not tested. The study only shows that there is a general mechanism in the brain responsible for the assessment of attractiveness. The fact that sexual preferences do not affect the response is not proof that attractiveness and reproduction are dissociated.

But generally, such an approach only sets a different level of argumentation. Even if beauty preferences themselves are an inherent result of our sensory and cognitive apparatus, this does not exclude the relation with the evolutionary principles of promotion of reproductive success and positive emotional feelings toward such stimuli. In this sense, we could argue that there is a co-evolution of evolutionary aesthetic principles and our perceptive apparatus.

6.5 Alternative Views: The Evolutionary Psychology of Ugliness

Grammer et al. (2001) tried to create attractiveness templates from body and facial features. The whole appearance appears to be coherent, and human bodies should be treated as a complete ornament, rather than an array of single traits (Thornhill & Grammer 1999). But the authors went one step further. Current literature describes beauty as a positive concept. However, a negative concept seems to work as well, if not better. If this is the case, research should consider redefining beauty as the 'avoidance of ugliness'. To test this hypothesis they correlated the attractiveness of the best single trait with male judgements. It shows that the 'single trait' method can predict male judgements. But the prediction is significantly more accurate when the worst trait is used. Such a method will only work when the traits of an individual are intercorrelated. This suggests a second dimension for beauty assessment in evolutionary aesthetics, namely the avoidance of ugliness. Indeed avoidance of stimuli could play a role as important as the positive appreciation of stimuli. Humans avoid stimuli like smells of skatole, or substances which taste bitter, so we may assume that these tendencies are innate and serve parasite and poison avoidance.

Di Dio et al. (2007) scanned classical and renaissance statues and altered their proportions. In an fMRI study, the unaltered statues showed an increased activation in the insulae which the altered failed to evoke. But when they asked the subjects for explicit beauty assessments, the beautiful images activated the emotional part of the brain, the amygdala. This evidence indicates that the brain reacts differently to beauty and ugliness, and is modulated by individual experience.

Experiencing beauty in art is the result of a neural agreement between two parts of the brain that govern the subjective and objective sides. On the basis of these results, we can suggest that there is also an inherent dimension of ugliness avoidance, which was probably generated during evolution.

6.6 The Future of the Adapted Mind

One feature of the adapted mind which is rarely addressed is the fact that any adaptation is prone to exploitation, and maybe even more dangerously, can be exploited by cultural developments. This does not imply that we are maladapted to the modern environment. It may simply be the case that new cultural developments can outrun biological adaptations. One example is the speed of information transfer. Many of our adaptations may be linked to slow and small information changes in our environment, while the invention of modern media techniques has created a completely new situation. Literature shows that beauty brings about status and success and — naturally unconsciously — reproductive success. If this is so, artificial body enhancements that amplify attractiveness will be widespread. As almost all cultures use such measures, this is nothing new. Yet this phenomenon is limited. By evolution we cannot become more beautiful, since high genetic variance is necessary for biological success. Consequently, if a certain beauty enhancement generates status, this enhancement will lose its advantage when used by too many people, and new enhancements have to be introduced. This is the eternal circle of new fashions and the invention of beauty products. Now this trickle-down mechanism meets prototyping. If beauty standards are a result of what people perceive in the mass media, exposure to media will change the prototypes. As the media themselves will use beauty for the status quest among different types of media, beauty standards will automatically trickle down in the media, and then a quest for more beauty will start. This is when plastic surgery and hormonal treatments come into play.

Beauty surgery, especially breast augmentation, plays an increasing role in today's societies. Surveys suggest that more than 800 000 American women have breast implants. The majority of such surgery is not motivated by medical reasons. Surveys suggest that the average woman desiring surgical breast augmentation is as psychologically stable as other women. In fact, such women differ only in small ways from other women, primarily in their negative evaluation of their breasts and their greater emphasis on dress and physical attractiveness (Shipley et al. 1977). Increased attractiveness is the core reason for plastic surgery. Indeed, in very young girls undergoing plastic breast surgery, the reduction of asymmetries accounts for more than 60% of cases (Grolleau et al. 1997). This correction would be expected, given the role of symmetry in beauty perception. In a study of women who have undergone breast augmentation, an interesting effect has been shown by Cook et al. (1997). Women with breast implants were more likely to drink a greater average number of alcoholic drinks, to be younger at first pregnancy, to be younger at first birth, to have a history of terminated pregnancies, to have used hair dyes, and to have had a greater lifetime number of sexual partners than women with no plastic surgery record. These differences between women with and without breast implants suggests that breast augmentation may lead to higher attractiveness for males, greater possible choice of high status males, and finally, greater possible reproductive success. But as soon as this circle is set up and success is triggered by surgery, its use will spread and trickle down to more and more surgery, until plastic people emerge.

That is the dark side to this game. Men who see movies with beautiful women adjust their beauty standards accordingly (Kenrick & Gutierres 1980; Kenrick et al. 1989). Then they develop greater aspirations for attractiveness in a dating experiment. The media can thus create 'unreal' beauty standards. When the media raise attractiveness standards by prototyping beauty, then unreal expectations for mate quality (beauty) will emerge. If the prototype is more beautiful than the mean in reality, no mate selection can occur on realistic grounds. One consequence of such a shift in attractiveness perception is the high number of singles we observe in Western societies.

The second, even more problematic development is coupled to the emergence of relatively new diseases like anorexia. Feminists blame social pressure for women's dissatisfaction with their bodies. In trying to achieve the slender, toned body that is associated with youth, women — and recently also men — even run the risk of developing eating disorders. Actually, this social pressure is caused by other women who compete for the same resources, just as much as by the beauty industry. It is a trickle-down phenomenon with a biological basis that has turned into an arms race of breathtaking speed with the help of the modern mass media. Will the outcome of this race be artificial people in artificial worlds?

References

Aharon I, Etcoff N, Ariely D, Chabris CF, O'Connor E, Breiter HC (2001) Beautiful faces have variable reward value: FMRI and behavioral evidence. Neuron 32(3):537–551
Baugh SG, Parry LE (1991) The relationship between physical attractiveness and grade-point average among college women. Journal of Social Behavior and Personality 6(2):219–228
Brown WM, Cronk L, Grochow K, Jacobson A, Liu CK, Popovi Z, Trivers R (2005) Dance reveals symmetry, especially in young men. Nature 438:1148–1150
Collins MA, Zebrowitz LA (1995) The contributions of appearance to occupational outcomes in civilian and military settings. Journal of Applied Social Psychology 25(2):129–163
Cook LS, Daling JR, Voigt LF, deHart P, Malone KE, Stanford JL, Weiss NS, Brinton LA, Gammon MD, Brogan D (1997) Characteristics of women with and without breast augmentation. Journal of the American Medical Association 277(20):1612–1617
Cosmides L, Tooby J (1992) Cognitive adaptations for social exchange. In Barkow JH, Cosmides L., Tooby J. (Eds.) The Adapted Mind. Evolutionary Psychology and the Generation of Culture (pp. 163–228). Oxford: Oxford University Press

Cunningham MR, Roberts AR, Wu CH, Barbee AP, Druen PB (1995) Their ideas of beauty are, on the whole, the same as ours — Consistency and variability in the cross-cultural perception of female physical attractiveness. Journal of Personality and Social Psychology 68(2):261–279

Darwin CR (1871) *The Descent of Man and Selection in Relation to Sex*. London: Murray

Di Dio C, Macaluso E, Rizzolatti G (2007) The Golden Beauty: brain response to Classical and Renaissance sculptures. PLoS one 2(11):e1201

Dion K, Berscheid E, Walster E (1972) What is beautiful is good. Journal of Personality and Social Psychology 24(3):285-290

Enquist M, Arak A, 1994. Symmetry, beauty, and evolution. Nature 372:169–172

Fink B, Grammer K, Mitteroecker P, Gunz P, Schaefer K, Bookstein FL, Manning JT (2005) Second to fourth digit ratio and face shape. Proceedings of the Royal Society B 272:1995–2001

Fink B, Grammer K, Matts PJ (2006) Visible skin color distribution plays a role in the perception of age, attractiveness, and health in female faces. Evolution and Human Behavior 27:433–442

Folstad I, Karter AJ (1992) Parasites, bright males, and the immunocompetence handicap. American Naturalist 139(3):603–622

Galton F (1879) Composite portraits, made by combining those of many different persons in a single resultant figure. Journal of the Anthropological Institute 8:132–144

Grammer K, Thornhill R (1994) Human facial attractiveness and sexual selection: The roles of averageness and symmetry. Journal of Comparative Psychology 108(3):233–242

Grammer K, Fink B, Juette A, Ronzal, G, Thornhill R (2001). Female faces and bodies: *n*-dimensional feature space and attractiveness. In Rhodes G, Zebrobwitz L (Eds.) *Advances in Visual Cognition. Volume I: Facial Attractiveness*. Ablex Publishing

Grammer K, Fink B, Møller AP, Thornhill R (2003) Darwinian aesthetics: Sexual selection and the biology of beauty. Biological Reviews 78(3):385–407

Grammer K, Keki V, Striebel B, Atmüller M, Fink B (2003). Bodies in motion: A window to the soul. In Voland E, Grammer K (Eds.) *Evolutionary Aesthetics* (pp. 295–324). Springer: Heidelberg, Berlin, New York

Grammer K, Fink B, Møller AP, Manning JT (2005).Physical attractiveness and health [comment on Weeden and Sabini (2005)]. Psychological Bulletin 131(5):658–661

Grogan SC (1999) *Body Image: Understanding Body Dissatisfaction in Men, Women, and Children*. London: Routledge

Grolleau JL, Pienkowski C, Chavoin JP, Costagliola M, Rochiccioli P (1997) Morphological anomalies of the breast in adolescent girls and their surgical correction. Archives de Pediatrie 4(12):1182–1191

Hamilton WD, Zuk M (1982) Heritable true fitness and bright birds — A role for parasites. Science 218(4570):384–387

Hatfield E, Sprecher S (1986) *Mirror, Mirror: The Importance of Looks in Everyday Life*. New York: SUNY Press

Hughes SM, Harrison MA, Gallup Jr. GG (2002) The sound of symmetry: Voice as a marker of developmental instability. Evolution and Human Behavior 23(3):173–180

Iliffe AH (1960) A study of preferences in feminine beauty. British Journal of Psychology 51(3):267–273

Ishai A (2006) Sex, beauty, and the orbitofrontal cortex. International Journal of Psychophysiology 63:181–185

Johnston VS, Hagel R, Franklin M, Fink B, Grammer K (2001) Male facial attractiveness: Evidence for hormone-mediated adaptive design. Evolution and Human Behavior 22(4):251–267

Kalkofen H, Müller A, Strack M (1990) Kant's 'facial aesthetics' and Galton's 'composite portraiture' — Are prototypes beautiful? In Halasz L (Ed.) *Proceedings of the XIth International Colloquium on Empirical Aesthetics* (IAEA). Budapest

Kenrick DT, Gutierres SE (1980). Contrast effects and judgements of physical attractiveness — When beauty becomes a social problem. Journal of Personality and Social Psychology 38(1):131–140

Kenrick DT, Gutierres SE, Goldberg LL (1989) Influence of popular erotica on judgements of strangers and mates. Journal of Experimental Social Psychology 25(2):159–167

Kranz F, Ishai A (2006) Race perception is modulated by sexual preference. Current Biology 16:63–68.

Langlois JH, Roggman LA (1990). Attractive faces are only average. Psychological Science 1(2):115–121

Langlois JH, Roggman LA, Rieserdanner LA (1990) Infants' differential social responses to attractive and unattractive faces. Developmental Psychology 26(1):153–159

Lorenz K (1973) *Die Rückseite des Spiegels*. Munich: Piper

Matts PJ, Fink B, Grammer K, Burquest M (2007) Color homogeneity and visual perception of age, health, and attractiveness of female facial skin. Journal of the American Academy of Dermatology 57(6):977–984

Møller AP (1992) Parasites differentially increase the degree of fluctuating asymmetry in secondary sexual characters. Journal of Evolutionary Biology 5(4):691–699

Mueller A (1993) Visuelle Prototypen und die physikalischen Dimensionen der Attraktivität. In Nikette R, Hassebrauck M (Eds) *Physische Attraktivität*. Göttingen: Hogrefe

Mueller U, Mazur A (1997) Facial dominance in *Homo sapiens* as honest signaling of male quality. Behavioral Ecology 8(5):569–579

O'Doherty J, Winston J, Critchley HD, Perrett D, Burt DM, Dolan RJ (2003) Beauty in a smile: The role of medial orbitofrontal cortex in facial attractiveness. Neuropsychologia 41:147–155

Penn DJ, Oberzaucher E, Grammer K, Fischer G, Soini HA, Wiesler D, Novotny MV, Dixon SJ, Xun Y, Brereton RG (2007) Individual and gender fingerprints in human body odour. Journal of the Royal Society Interface 4(13):331–340

Perrett DI, May K, Yoshikawa S (1994) Facial shape and judgement of female attractiveness. Nature 368(6468):239–242

Rikowski A, Grammer K (1999) Human body odour, symmetry, and attractiveness. Proceedings of the Royal Society London B 266:869–874

Roberts SC, Lile AC, Gosling LM, Perrett DI, Carter V, Jones BC, Penton-Voak I, Petrie M (2005) MHC-heterozygosity and human facial attractiveness. Evolution and Human Behavior 26(3):213–226

Schaefer K, Fink B, Grammer K, Mitteroecker P, Gunz P (2006) Female appearance: Facial and bodily attractiveness as shape. Psychology Science 48(2):187–204

Shipley RH, Odonnell JM, Bader KF (1977) Personality characteristics of women seeking breast augmentation — Comparison to small-busted and average-busted controls. Plastic and Reconstructive Surgery 60(3):369–376

Thornhill R (1992) Fluctuating asymmetry and the mating system of the Japanese scorpion fly *Panorpa japonica*. Animal Behaviour 44(5):867–879

Thornhill R, Grammer K. (1999) The body and face of woman: One ornament that signals quality? Evolution and Human Behavior 20:105–120

Van Valen L (1973) A new evolutionary law. Evolutionary Theory 1:1–30

Voland E, Grammer, K (Eds) (2003) *Evolutionary Aesthetics*. Springer: Heidelberg, Berlin, New York

Zahavi A, Zahavi A (1997) *The Handicap Principle: A Missing Piece of Darwin's Puzzle*. New York, Oxford: Oxford University Press

Chapter 7
Our Appetite for Information: Invented Environment, Non-Transparent Mind, and Evolved Preferences

Matthias Uhl

Abstract Evolution and its shaping of our minds are in part responsible for our interest in the media, especially its fictional content. This statement does not mean that everything in the media and fictional plots can be explained on a biological basis. The two other important ingredients are of course the socialization of the individual and the culture it takes place in. The three factors interact.

7.1 Introduction

This is the age of media and information. Today's technologically enhanced communications seem to be light-years away from the times when there was no reading and writing, and not even language as we know it today. But the idea that 'everything is new' when we look at cell phones, HD-TV, and internet is quite wrong. It is wrong because these high-tech products just provide a new playground for old psychological mechanisms. The human brain has been shaped by the long road of evolution that led from small nocturnal mammals 65 million years ago to today's couch potatoes, online gamers, and movie freaks. And the imprint of evolution has severe consequences for the way people perceive and use the media.

In the following, I will highlight five evolutionary insights into the relationship between man and media:

- Our ability to deal with the media rests on evolved mechanisms for stimulus perception and processing.
- The mechanisms involved are only in part accessible to our consciousness.
- The psychological mechanism of attention — our guide through media content — is shaped by the environments we evolved in.

Matthias Uhl
Phönixstr. 3, 35578 Wetzlar, Germany, e-mail: m@tthiasuhl.de

U.J. Frey et al. (eds.), *Essential Building Blocks of Human Nature*, The Frontiers
Collection, DOI 10.1007/978-3-642-13968-0_7, © Springer-Verlag Berlin Heidelberg 2011

- It turns out to be an illusion that the perception of media is somehow separated from our perception of the 'real world'.
- Our minds integrate both of these worlds and produce strange effects like fictional character friends and an altered perception of the dangers of everyday life.

7.2 Old Cognition, New Playgrounds, and the Technological Fallacy

Have you ever met someone who thinks that virtual reality is one of the coolest things on earth, or will be when the equipment gets better? Being able to enter artificial 3D worlds seems to be the high point of media technology, maybe even the next level of culture. Well, if you think so, you are a victim of what I call the technological fallacy. Why is this? Because even rodents have no problem at all handling an artificial 3D environment. Experiments by Hölscher et al. (2005) have shown that rats learn to handle virtual spaces in the same way as humans. The animals were surrounded by a 360-degree screen and placed on a movable ground, which transformed their running into computer input for adjusting the virtual environment to their actions. The world they found themselves in was not very sophisticated. It was a broad plain with big objects hanging from the ceiling every two meters. Whenever the rats crossed the area directly under such an object they received a reward in the form of a drop of sugar water in their mouth. Having a sweet tooth, the rats soon began to optimize their patterns of movement to maximize the sugar harvest, a development very similar to the learning behavior you can see with computer gamers.

What I mean by technological fallacy, when observing these clever rats, is that humans tend to mistake the technological difficulty of a medium for the cognitive complexity required to deal with it. Interactive virtual reality first emerged in research labs in the last decade of the last millennium. But the cognitive apparatus required to deal with it by integrating visual input and motor action seems to date back to the beginning of the rise of mammals 65 million years ago (Zeki 1999). If rodents can use virtual reality so easily, it cannot be such a big cultural issue on the part of the consumer. The term 'culture' refers to all kind of abilities that humans learn during their individual development. Now it definitely is very elaborate culture on the part of the scientists and engineers who make it all work. But just being able to immerse oneself in such a world is not.

You probably know the stickers that often come with new computers: "Ready for [...]". If your brain had had a similar sticker at birth, it would have read "Ready for virtual reality". The third of your brain located in the back of your head is made just for this, rendering 3D pictures in real time, in millions of colors, and under any lighting conditions. But if the sticker had read: "Ready for reading", after a few years, you would definitely have looked for someone to sue. The reason is obvious. Your neurobiological equipment is not at all ready to read when you grow up. Of course, you can learn to do it, but it takes hundreds and thousands of hours. And

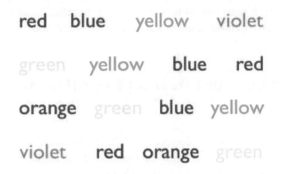

Fig. 7.1 Naming the colours of words

more importantly, by this intensive training, you develop an ability that not only exploits your neuro-cognitive mechanisms, but actually alters them. Have a look at Fig. 7.1 and try to name as quickly as possible the colors the words are written in.

Trying to name the word's color rather than the color it refers to has probably resulted in a bit of stuttering. The color the word signifies nearly always gets in conflict with the color the word is written in. It takes some concentration to stick to the right kind of colors. If you challenge an illiterate with the same task, he will have no problem at all. The difference between the two of you is your brain. These endless hours you have spent reading have shaped connections between nerve cells that compulsively transform groups of letters into meaning. You have worked so hard on it that you may never before have realized that you cannot stop this permanent production of meaning when written words come into focus. This is what happens when you look at the colored color word. Your color vision provides you with one result for the question you are trying to answer, while your language center comes up with a different answer. The stuttering occurs because not only the desired answer makes its way into your consciousness, but the other one tries to as well. The unconscious parts of your brain deliver two answers, and you have to decide which one to take.

If you define culture as a set of abilities that can be transferred between humans via learning, the difference between virtual reality and reading is now clearly visible. On the consumer side, and that is what I am talking about here, virtual technologies utilize the cognitive mechanisms fundamental to everyday life. Reading on the other hand is only possible after a long phase of training that leads to permanent changes in nerve connections. No learning is needed to perceive an artificial 3D picture, but learning is unavoidable if you want to read and understand this sentence.

The cultural investment for handling these technologies — we are still on the consumer side — is far higher when it comes to reading than when it comes to entering virtual worlds. But when you look at the technologies, the simple appearance of a book seems a bit outdated compared with headsets, controllers, and so on. And this difference in user interface often leads to the impression that virtuality is the 'wow-thing'. Well, it may indeed be a wow-thing, but that is not because of a new level of culture. It is because of an evolutionarily very old one.

7.3 We Don't Know that We Don't Know How We Perceive Media

The experiment and the experience provided in the lab were definitely both rather strange. A man was sitting on a chair wearing a head-mounted stereo display (Ehrsson 2007). Two meters behind his back was a camera taking a 3D picture of this perspective and sending it directly into the display. So the man on the chair was looking at his own back. Then the experimenter touched the participant's chest, and also the area below the camera lens. The camera picture clearly showed a hand with a plastic rod moving towards the chest of the turned-away body. Simultaneously the picture showed a second hand with plastic rod passing through the lower part of the visual field, in just the direction of the chest, as if this were the view from some real eyes.

Being part of this very artificial arrangement had a nearly metaphysical effect on the participants. What their senses and brains made of this situation is usually known as an out-of-body-experience. They felt that they were looking at their own body while being located two steps behind it — just the same phenomenon known from some cases of near-death experiences. When asked about the experiment, most of the participants said "I experienced that I was located at some distance behind the image of myself, almost as if I was looking at someone else" and "I experienced that the hand I was seeing approaching the cameras was directly touching my chest (with the rod)". Ehrsson saw this as clear evidence for the hypothesis that our self, or the I-perspective, is not slavishly fixed to our body, but can be manipulated.

There is a second conclusion that can be drawn from this experiment: the media influence humans in ways that are not at all obvious, and may even be quite unexpected. But the same people who admit that these results are really stunning usually deny that their behavior is influenced by their media consumption. If this were the case, they would know, so the reasoning goes. Let us call this assumption the self-transparency hypothesis. It says that people are fully aware of what the media do to them because they consciously experience and process their media consumption.

This is a very plausible argument, but it nonetheless rests on a hidden and quite problematic assumption. What if the processing of media stimuli coming to our senses is not fully conscious, or in part cannot even be reached by our consciousness? The feeling that we definitely know what is going on in our head is strong, but wrong. The reason for this mistake lies in the way our brain has been shaped by evolution. If you put your hands together as if you were about to carry some water and now place them on your forehead, that is where your consciousness sits. The neocortex, the outer part of the brain in this region, is the tissue that generates the states of mind we experience. If you put your hands on the back of your head, that is where visual processing takes place. The rear third of your brain is constantly oc-

Fig. 7.2 Spot the triangle

cupied transforming the input of the one million nerve fibers coming from each eye into a real time 3D picture of the world. Only when this picture is finished does it get forwarded to the frontal regions of the brain where we get the chance to experience it consciously.

Let us carry out a short experiment to get a better understanding of this non-conscious processing. You have probably already seen triangles like the one shown in Fig. 7.2. It is not a complete geometric figure. The corners are there but the edges in-between are missing. Looking at it, you still get the impression that there are something like edges on the white paper. Of course, you know that there are not. This is a widely known optical illusion.

The most popular explanation for this false impression is that we are somehow trained to recognize triangles. Think of the endless hours of geometry you have had in school. Maybe as a result of all these exercises your brain developed the habit of automatically completing triangles. This is a charming just-so-story. Your seeing edges where there aren't any has nothing to do with your educational career. People who have never seen a school from the inside have just the same impression as you do, and even monkeys join in for this experience. How do we know? The visual cortex in the back of a monkey's brain is structured in just the same way as ours. One of the areas within this occipital lobe is specialized in edge detection. If there is an edge in part of the visual field, a nerve column is responsible for the highly active processing nerves communicating their finding to higher visual areas. If the edge vanishes, the activity ends. Visual illusions like the triangle above induce a third state, namely, a 50 percent activity that somehow leads to the strange impression of seeing an edge where there definitely is none (Hoffmann 2000). The same process that was measured in monkey brains is taking place in your head when you look at this kind of illusion.

The auto-completion mechanism triggered by this triangle is often explained by referring to the relationship between early humans and tigers. Being unable to react adequately when seeing just a tiger's tail from behind a bush was a pretty bad thing in terms of survival. Seeing the whole tiger a little bit later was probably too late to save your genes for the next generation.

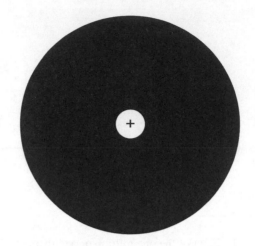

Fig. 7.3 Seeing circles where there are none

This tiger story gets the point. Our perception, and even more so our action planning, is not made in such a way as to wait until the uptake of information is completed. Put philosophically, you do not need to know the truth, you just need to know enough to make the right decision. Acting quickly and in approximately the right way is nearly always better than acting slowly and with absolute certainty.

Let us do a second experiment. In Fig. 7.3 you see a massive circle with a cross in the middle. After you have read this paragraph you should give this a try: look at it for approximately a minute, focussing on the cross and making sure that your eyes do not wander around. After staring for a minute, you then look at a white wall or a white sheet of paper in front of you. Looking at both of them successively is also a way to enjoy the effect that should be present in your visual field.

If you have looked at the cross in the middle of the circle for long enough and than turn your eyes away, there should be a white circle either on the wall or on the sheet of paper in front of you. This circle persists for quite a while. If you go from looking at the wall to looking at the white sheet, you will notice that the size of the circle changes. Depending on how far away the wall was, the diameter of the circle might have reached a meter or more. When looking at something white directly on the table in front of you, this measure shrinks to a little bit more than 10 centimeters.

What is happening here and why is it important in relation to media perception? What you experience is an effect of modified sensitivity in the sensory cells of your retina. The cells producing the visual impression of the black circle are relaxing in a certain sense. By not being exposed to light, they build up the pigment rhodopsin and thus became more sensitive to light. When looking at a white surface afterwards, they produce the impression of a whiter white than the non-regenerated neighboring cells.

That is the explanation of the effect. This experiment is so interesting in relation to the media because it demonstrates that, for at least a part of what we perceive, we are not aware of the processes involved. That is not surprising at all. Try to think as deeply as possible. You will never reach a point allowing you to realize that all your thoughts are the consequences of neurons interacting. If you want to use an analogy from computer science, your mind provides you with a user interface that hides the actual mechanisms doing the work. Nevertheless, we tend to think that we are fully aware of what is going on.

7.4 Still the Same Old Cues

As Anne Campbell has phrased it, evolution is "random genetic variation, non-random selection" (Campbell 2000). Organisms adapt to their environment by this mechanism. Looking at primates, it is not surprising that not only the body adapted, but the senses and their processing as well. Detecting edges and changing sensitivity are useful mechanisms in everyday life and can be deployed when using media as well. The adaptations that developed in the past did not stop there. Not all parts of a primate's environment are equally important for its survival. The other members of the group are the key factors in this game. The psychological mechanism of attention shows clear evidence for this.

In a very sophisticated experiment, Robert Deaner and colleagues (2005) showed that rhesus macaques are willing to pay (orange juice) for the possibility to look at pictures of members of their own group. But the attention given to the photos was not evenly distributed. No one was willing to pay for the lower end of the hierarchy, but everyone paid to see the top dogs. This result makes perfect sense when you consider a strategy for getting along in a group of monkeys. Your life pretty much depends on the behavior of the leader(s), not on what the losers do. But why do these monkeys display the same behavior when looking only at pictures? The boss will never know or even appreciate how much they are willing to spend to look at a technical reproduction of his face. In the end, they just seem to have wasted limited resources to achieve nothing. The reason for this irrational behavior is that the attention mechanism is not able to draw a clear distinction between real world and media stimulus. The alpha's face draws attention wherever it appears. In a sense, the medium misuses this weakness.

The same is true for humans. Faces are an important source of information for a socially living species like our own. Knowing this, it is not surprising that evolution has equipped us with a natural attraction towards faces. Looking at a scene from *Who Is Afraid of Virginia Woolf*, Ami Klin (2002) showed how spectators use this part of the body while trying to make sense of an interaction. When the visitor asks his host about the painter of one of the pictures on the wall, the eyes of the viewer start on the visitor's face, follow his pointing towards the picture, and from there move on to the host's face rather than return to the face of the visitor. Klin used a device called an eye tracker to pinpoint the exact route of the onlooker's attention. This is possible because our eyes have just a small area of very detailed vision that

Fig. 7.4 The importance of face recognition

is constantly adjusted to a point of maximum attention in our environment (Gregory 1997).

The innovative element of Klin's study is his choice of participants. Beside normal viewers, he had a highly functional autistic person watch the movie while monitoring his gaze. 'Highly functional' means that the participant had a university degree. The core element of autism is the absence of a theory of mind (Cohen 2003). People with this syndrome are unable to access the thoughts and feelings others might have. To be usable, information has to be explicit. The only explicit information in the mentioned sequence was the guest asking for the artist who made one of the pictures on the wall. The focus of this participant started somewhere below the speaker's face and moved to the wall without using the arm gesture as an indicator. Not being able to identify the right picture, wandering from left to right, the eye focus remained on the wall when the normal viewer had already completed his circle of attention.

We are usually not aware of our special fondness for faces. Non-autistic humans have an area in their visual cortex that they use especially for analyzing faces. It is like a microscope that always kicks in when we look at faces and gives us a huge amount of extra information. If this area is damaged, life gets very complicated because all people with roughly the same haircut and color look alike.

You can experience what a difference it makes depending on whether this face detection area is involved or not. Have a look at the photo in Fig. 7.4 and experience the so-called Thatcher effect. You see me (I do not have the copyright for the original photo) upside-down. You will realize that there is something wrong with my face.

The mouth and the eyes seem to be modified somehow. Now turn the book upside down and look at me again.

Your vague notion was right: the mouth and eyes look grotesque. They are turned upside down in the otherwise unchanged face. As long as you look at it while holding the book the ordinary way, it is a little strange. Upside down I look like a monster. This difference is caused by the activity of the face-analyzing region in your brain, which starts to work as soon as the orientation of my head is close to normal. Your may not realize it, but the unconscious part of your brain has clear preferences when looking at the world.

7.5 Reality and Media: The Boundary that Never Was

In the 1980s, George Gerbner and colleagues looked at TV viewers and discovered an effect they called mainstreaming (Gerbner 1989). They found that the time spent in front of the television was inversely proportional to the diversity of opinions the viewers possessed. The longer people watched every day, the closer their views of the world were, no matter what their social background.

This is interesting, but somehow expectable. The unexpected finding was that more TV consumption led to a feeling of higher risk when outside. The more people watched, the more dangerous their environment appeared to them. But why should this be so?

Let us look at another experiment that may help to solve this puzzle. Byron and Reeves (1998) placed participants in front of a computer and had them solve some tasks. At the end, a questionnaire appeared on the screen asking how helpful the machine had been. The control group solved the same tasks, but then had to change rooms to fill out the final questionnaire on another computer. The difference between these groups was significant in a strange way. People who did the final evaluation on the same computer as the previous work rated its usefulness much more positively. After running several other experiments to check their hypothesis, Byron and Reeves ended up with just one explanation for their results. The participants were polite when answering on the same machine and honest when using the other one. Of course, they were not aware of being polite — they strictly denied this possibility when interviewed — but their behavior was indeed more polite. The people were treating the machine like a human without realizing it.

The reason for this behavior lies in what Pascal Boyer calls ontological categories (Boyer 2002). During human evolution, our psyche developed a simple but efficient categorization for two different classes of objects in our environment: physical objects and protagonists, meaning all kinds of living creatures — including humans — that follow their own interests by more or less complex actions. This second category of entities in our environment requires much more attention and different handling than mere physical objects. If you meet someone again, he will probably treat you better when you said something nice the last time you met. This is exactly what we humans do, but unconsciously, as the experiment proved. The clear cut line that we draw between humans and machines does not go along with our behavioral preferences. If things, e.g., computers, interact with us in ways that for millions of years were exclusive to human interaction, we have a tendency to treat them like humans. Of course, we know that we are dealing with a machine, but our evolved cognitive mechanisms make us — without realizing — treat it differently to a simple physical object.

Returning to George Gerbner's high intensity TV users, these people felt more afraid of being robbed or murdered in the streets than their more moderate TV watching neighbors. The reason seems to be an error in the cognitive separation of information perceived from the real world environment and the consumed media content. Of course, these people know that they are watching TV. But this is just the result of what the conscious part of the brain says. Somewhere in the unconscious remainder of the brain, the information about witnessed human interactions gets integrated together, mashing up reality and fictional material. The subconscious mechanisms that monitor our environment did not evolve for a world with technologically enabled mass communications. For more than 99 percent of history and prehistory, all perceivable human interactions were directly relevant for the individual witnessing them, because they necessarily happened nearby. From the standpoint of cognitive evolution, it made perfect sense to monitor them constantly and then transform them into a feeling of how safe or dangerous a place or person might be.

The advent of the media has slowly changed this situation. These means of communication have made it possible to participate in remote or even fictional interactions. The screen media enhanced this experience by providing a catchy unity of motion picture and sound. But still these non-present or even non-real interactions continue to influence the cognitive mechanisms evaluating our environment. This is the reason behind Gerbner's findings. If you doubt this explanation, try taking a walk in a forest by night just after you have seen a comedy or romance and compare it with the same experience just after you have seen a horror movie. You will definitely notice that the different media consumption changes the way you feel in the dark below those old trees. The media content we consume modifies our emotional response to the environment.

7.6 The Old Stories Are the Best

Up to now we have taken evolutionary thinking quite some way into the realm of the media. Due to the structure of our neural processes, we are unable to perceive the ways our cognition handles such stimuli. And neither are we aware that our attention is not evenly distributed throughout the world, but has hotspots of special interest. These hotspots, like the human face, developed during human evolution as part of strategic information management.

But one can take these thoughts even further. Back in 1949, Joseph Campbell published his book *The Hero with a Thousand Faces* (Campbell 1949), in which he demonstrates that mythological hero stories from all around the world have a

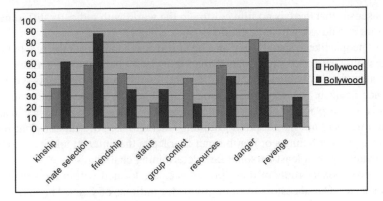

Fig. 7.5 Comparing the importance of the main issues in Hollywood and Bollywood films

common plot structure. Joseph Carroll in his study *Evolution and Literary Theory* (Carroll 1994) made the connection with biology and suggested that the basic plots in literature have their roots in our evolutionarily-shaped psychology. Ed Tan (Tan 1996) and Thorben Grodall (Grodall 1997) separately developed very similar considerations when looking at films. Movies fulfill needs that are not only generated by socialization but also by much older preferences. Clemens Schwender, analyzing media in general, compressed this phenomenon in the statement that the media act as dummies for our cognition.

In 2002, using this as a starting point, Peter Heyl, Manfred Kammer, Keval Kumar, and myself began to compare Hollywood and Bollywood movies (Uhl & Kumar 2004, Uhl & Hejl 2010). The criterion for choosing the films was financial success. We gathered the top hundred movies from each of these film production centers, the biggest in the world, and analyzed their content and plot structure.

Looking for the main elements that drive the story of a movie, we found eight core factors: kinship, mate selection, friendship, status, group conflict, resources, danger, and revenge. We developed a questionnaire that media students used to analyze the movies. The graph in Fig. 7.5 shows the results on the top level, comparing the importance of the main issues in Hollywood and Bollywood films.

The main result is that there is no qualitative difference. The films of both cultures utilize the same elements to construct their plots. A closer look shows that the ranking and frequency of these elements is different in a way that can be connected to the cultural context. Indian films reflect the importance that is given to family and marriage within their cultural environment. Western films show a stronger emphasis on problems which individuals or groups of individuals have to face. The obvious differences between the standard Western and the standard Indian film are its length and the importance of song and dance. Indian films usually take twice as much time to tell their story. They also include five to eight song and dance sequences, more or less well integrated into the development of the story. We discussed these results with specialists for African, European, South American, and Asian film and reached the consensus that there is no film culture in the world with radically different plots. Again we should ask why this should be so.

Most people like to spend a day on a beautiful beach looking at the sea, but almost no one would pay to see such a panorama for one and a half hours in a movie theater. The reason for this behavior, and also for the similarities in film plots worldwide, lies once again in the way information is managed by our evolved psyche. The beach is a nice place to be, but not an interesting place to look at, because most of the time there are no attention-grabbing interactions. And this is what humans want to see: other humans or non-human characters that interact and have to handle life-threatening or at least personal fate-determining challenges.

The evolutionary roots of these preferences are located within the social nature of our species. Our ancestors lived in groups for millions of years. The narrow focus of primate or protohuman attention did not allow constant monitoring of all ongoing interactions, whence a preference developed for viewing strategically valuable interactions. But the strategically valuable interactions are especially those that change the existing hierarchy: who moves upwards, who moves down, which new alliances form, who gets into conflict, and how do the involved individuals perform? Movies deliver the kind of information about their characters that you would need to know if they were part of your real environment. Looking at someone sleeping tells you nothing. Watching a short fight may influence all your planning for the future.

7.7 Conclusion

Evolution and its shaping of our minds are in part responsible for our interest in media, especially its fictional content. But this statement does not imply that everything in media and fictional plots can be explained on a biological basis. The two other important factors are of course the socialization of the individual and the culture it takes place in. And the three factors interact.

Biology does not explain why the Danes invented the dogma film style, why Indian films must have song and dance, and why Hollywood blockbusters try to use Joseph Campbell's analysis of myths as a blueprint. Evolutionary thinking tells us instead why we want to see people competing for scarce resources, why we sometimes want to see characters threatened with death, and why we even contemplate stories we would hate to be a part of. All these adventures are treasure troves filled with information about characters who, through the medium, seem to be close by (Gottschall & Wilson 2005). Not being equipped to draw a clear line between real and fictional action, our brain gets sucked in by these invented worlds (Uhl 2009).

It may appear to be a weakness of our minds to be attracted by the media and fictional content. But without these evolved preferences, there would be no theater, literature, and feature films at all. Our biology does not explain the whole realm of media and fiction, but it explains why we take these cultural inventions so seriously, and sometimes even get the feeling that their content affects our mortal existence within an ever-evolving world.

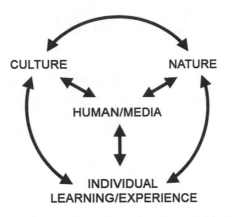

Fig. 7.6 Interactions between humans, the media, and environmental factors

References

Boyer P (2002) *Religion Explained*. Publisher?
Byron R, Nass C (1998) *The Media Equation*. Stanford, CSLI
Campbell A (2002) *A Mind of Her Own. The Evolutionary Psychology of Women*. Oxford, Oxford
    University Press
Campbell J (1949) *The Hero with a Thousand Faces*. New York, Pantheon Books
Carroll J (1994) *Evolution and Literary Theory*. Columbia, University of Missouri Press
Cohen SB (2003) *The Essential Difference. Men, Women, and the Extreme Male Brain*. New York,
    Basic Books
Deaner R, Khera A, Platt M (2005) Monkeys pay per view: Adaptive valuation of social images
    by rhesus macaques. Current Biology 15:543–548
Ehrsson HH (2007) The experimental induction of out-of-body experiences.
    Science 317(5841):1048
Gerbner G et al. (1989) The 'Mainstreaming' of America. Violence Profile No. 11. Journal of
    Communication 30(3):10–29
Gottschall J, Wilson SD (2005) *The Literary Animal. Evolution and the Nature of Narrative*.
    Evanston, Northwestern University Press
Gregory RL (1997) *Eye and Brain: The Psychology of Seeing*. Princeton, Princeton University
    Press
Grodall T (1997) *Moving Pictures. A New Theory of Film. Genres, Feelings, and Cognition*. Ox-
    ford, Clarendon Press
Hoffman DD (2000) *Visual Intelligence: How We Create What We See*. New York, Norton
Hölscher C, Schnee A, Dahmen H, Setia L, Mallot HA (2005) Rats are able to navigate in virtual
    environments. Journal of Experimental Biology 208(3):561–9
Klin A, Jones W, Schultz R, Volkmar F, Cohen D (2002) Defining and quantifying the social
    phenotype in autism. The American Journal of Psychiatry 159(6):895–908
Tan E (1996) *Emotion and the Structure of Narrative Film. Film as an Emotion Machine*. Mahwah,
    Erlbaum
Uhl M, Kumar K (2004) Indischer Film. Bielefeld, transcript
Uhl M (2009) Medien — Gehirn — Evolution. Mensch und Medienkultur verstehen. Eine trans-
    disziplinäre Medienanthropologie. Bielefeld, transcript
Uhl M, Hejl P (2010) Die alten Geschichten sind die Besten. Eine evolutionstheoretisch-inhalts-
    analytisch vergleichende Untersuchung westlicher und indischer Erfolgsfilme. In Buck M,
    Hartling F, Pfau S (eds) *Randgänge der Mediengeschichte*. Wiesbaden, VS-Verlag
Zeki S (1999) *Inner Vision*. Oxford, Oxford University Press

Chapter 8
Our Best Shot at Truth: Why Humans Evolved Mathematical Abilities

Niklas Krebs

Abstract This chapter discusses the evolutionary origins of the mathematical abilities of modern humans. I begin by analyzing what is actually meant by mathematical abilities and how they can be approached by considering a brain that is organized in a largely modular way. Emphasis is given to the analysis of the ontogenetic and phylogenetic development stages of the individual mathematical abilities. These individual aspects are then examined in more detail with regard to their evolutionary origins, discussing the question of whether or not they also possess an evolutionary function. Four hypotheses will be advocated. The first is that number sense is a module in human and animal brains which has the evolutionary function of being able to approximately grasp and process the quantity of elements in a given set. The second claims that the processing of quantities in symbolic form is a byproduct of a cognitive adaptation, the understanding of symbols. Thirdly, mathematical thinking in terms of relations is a byproduct of complex social thinking. Fourthly, there is no direct linear line of development in either the ontogenetic or the phylogenetic development of the mathematical abilities of modern humans.

8.1 Introduction

When contemplating the mathematical abilities of modern humans (*Homo sapiens*), one immediately thinks of natural numbers 1, 2, 3, and so on. The ability to count appears to be a very fundamental characteristic. But is this really true? Cannot other species count as well? Or is the ability to count a more highly developed cognitive ability which precedes other abilities, such as grasping the quantity of elements in a given set? And what about higher mathematics, which goes far beyond the ability to

Niklas Krebs
University of Giessen, Zentrum für Philosophie und Grundlagen der Wissenschaft, Otto-Behaghel-Str. 10C, 35394 Giessen, Germany, e-mail: Krebs-Homebase@t-online.de

count? In other words, the ability to count appears to be only one link in the chain of cognitive abilities that have evolved during the course of hominization.

In view of the foregoing, the questions that arise are, on the one hand, those concerning the evolutionary roots of these mathematical abilities, and on the other, the question of the ontogenetic and phylogenetic development processes regarding these mathematical abilities. To be able to discuss these questions, it will be necessary first to clarify what is meant by the mathematical abilities of modern humans. One must then examine the possible cognitive correlations of these individual components, reviewing possible evolutionary origins and then analyzing the evolutionary functions they may have. In particular, one must investigate the possible evolutionary correlations of the individual components of this essential building block of human nature, before attempting a discussion of the mathematical nature of modern humans.

8.2 The Mathematical Abilities of Modern Humans

What do we mean by the mathematical abilities of modern humans? To begin with, the list should include the cognitive ability to approximately grasp and process the quantity of elements in a given set. This cognitive ability is called number sense (Dehaene 1997, pp. 4–5) or number module (Butterworth 1999, pp. 6–10).

8.2.1 Number Sense

Number sense — an accumulator mechanism — is a peripheral module occurring equally in the brains of animals and humans, which facilitates the idea of quantity and its transformation according to the elementary rules of arithmetic (Dehaene 1997, pp. 4–5). It has already been demonstrated for many species, e.g., rats (Meck & Church 1983), parrots (Pepperberg 1987), tamarins (Hauser et al. 2003), rhesus monkeys (Brannon & Terrace 2000; Cantlon & Brannon 2007; Hauser et al. 2000; Washburn & Rumbaugh 1991), orang utans (Call 2000), chimpanzees (Boysen & Berntson 1989; Matsuzawa 1985; Woodruff & Premack 1981), human children (Lipton & Spelke 2003; Wynn 1990, 1992, 1995, 1998, 1998a; Xu et al. 2005) and adults (Barth et al. 2003, 2006; Dehaene 1997, pp. 64–72; Whalen et al. 1999). For a better understanding of what precisely is meant by the functioning of this number sense, the works of Pepperberg (1987) and Boysen and Berntson (1989) will be briefly discussed.

Pepperberg (1987) trained an African gray parrot to answer the question 'How many?' in verbal–numerical terms. Quantities of from two to six objects (e.g., little pieces of wood) were presented to the parrot and, in view of the positive results, Pepperberg came to the conclusion that limited, but nevertheless numerical abilities can be demonstrated in gray parrots. The questions of whether one can speak of the ability to count in this case or whether these proven numerical abilities can develop into the ability to count in gray parrots were left unanswered by Pepperberg. What is certain here at least is the existence of the accumulator mechanism, namely number sense.

On the other hand, Boysen and Berntson (1989) trained a 5.8 year old female chimpanzee (*Pan troglodytes*) to use Arabic numerals (0, 1, 2, 3, 4) in order to express the quantity of a given set of pieces of food. And conversely, the chimpanzee was able to select the corresponding set of pieces of food for a given Arabic numeral (0, 1, 2, 3, 4). In summary, they (Boysen & Berntson 1989) came to the following conclusion:

> These findings demonstrate that counting strategies and the representational use of numbers lie within the cognitive domain of the chimpanzee and compare favorably with the spontaneous use of addition algorithms in preschool children.

In other words, this positive result once again demonstrates the existence of an accumulator mechanism, i.e., number sense, although even more was proven here: not only the grasp and processing of quantities, but also its symbolic representation in the form of Arabic numerals.

In her review article on the empirical findings to date concerning young human children, Wynn (1998) also pursued the issue of where the evolutionary origins of numerical skills were to be found. She also argued in favor of an evolved mental mechanism for quantities, i.e., an accumulator mechanism, which humans share with other species and which represents the cognitive basis for all the more advanced mathematical abilities of human beings (Wynn 1998):

> Rather, findings support the existence of a dedicated mental mechanism specific to number, one which may have evolved through natural selection and which we may share with other warm-blooded vertebrate species. This mechanism serves as a foundational core of numerical knowledge, providing us with a toe-hold upon which to enter the realm of mathematical thought. But it is strictly limited in the number values it can represent and in the kind of numerical knowledge it supports.

The interesting aspect of these positive findings concerning the number sense is that especially adult chimpanzees and human children under the age of four years can apparently not only approximately grasp and process quantities, but can also prove this grasped quantity with a cardinal number reflecting the quantity of elements in the given set (e.g., of pieces of food). This basically means that numerical symbols in the form of Arabic numerals are also being comprehended and processed (at least approximately).

This is a cognitive ability which goes beyond the mere collection and processing of a quantity. Although number sense is an accumulator mechanism, it is unable to count a given set of objects discretely by means of Arabic numerals, for example. On the one hand, it is much too imprecise for this, as shown by the frequency of errors with increasing counts in the various experiments (e.g., Dehaene 1997, pp. 66–72) and on the other, it gets by without language (e.g., Dehaene 1997, pp. 86–88; Dehaene et al. 1990; Gelman & Butterworth 2005). Number sense estimates the quantity and processes this approximately, i.e., without representing it symbolically and without counting!

This ability to grasp a quantity approximately without counting is called subitizing in psychology (e.g., Dehaene 1997, p. 68; Kaufmann et al. 1949). In the case of adult humans, subitizing functions readily up to the number 3 and only as of the number 4 do the frequency of errors and the reaction time increase (e.g., Dehaene 1997, Figure 3.2; Mandler & Shebo 1982). The question that now arises is how the transition occurs from number sense, i.e., the accumulator mechanism which approximately grasps and processes the quantity, to symbolic number processing in the form of Arabic numerals, for example.

8.2.2 Number Processing

What the study by Boysen and Berntson (1989) made clear was that the numerical understanding of chimpanzees (*Pan troglodytes*) goes beyond mere number sense, as this study demonstrated that chimpanzees were even able to relate a number of pieces of food to the corresponding Arabic numeral. This does not necessarily mean that chimpanzees are able to count, i.e., have an understanding of the numerical relationships held by numerical symbols in the form of Arabic numerals. However, chimpanzees nevertheless have to have understood the principle that the approximate quantity of objects in a given set can be accounted for by a numeral, which in turn then represents an approximate number of objects. On the other hand, the question of whether chimpanzees can discretely count a given quantity of objects by means of learned Arabic numerals, in which case they would also have developed an understanding of the fact that the last number symbol when counting represents the number of objects of a given set, was definitely not demonstrated in this study and remains open.

In the case of human children, Wynn (1990) examined this so-called cardinal word principle, and came to the conclusion that only children from the age of 3.5 years are capable of this (Wynn 1990, Fig. 2). What is interesting about this result is the age of the children, in particular because it coincides with the occurrence of another cognitive ability, namely theory of mind (ToM). An individual is said to possess a theory of mind if it is able to ascribe mental states to itself and to others (e.g., Leslie 1987; Perner 1991; Premack & Woodruff 1978; Whiten 1996). Interestingly enough, until today, there has been no real agreement on whether chimpanzees and other great apes have a fully developed ToM, as their mental abilities in this respect correspond to those of children just under the age of 3.5 years (e.g., Call & Tomasello 2008; Suddendorf & Whiten 2003; Whiten 1996). So it could definitely be possible that an individual needs a ToM to be able to understand the cardinal word principle, or the cardinal number principle if the linguistic aspect is ignored. This would also explain why human children from an age of just over 3.5 years are capable of doing so relatively suddenly, whereas children under the age of 3.5 years and adult chimpanzees are not yet able to demonstrate such conduct.

Nevertheless, as explained above, the symbolic understanding of numbers by chimpanzees is sufficient to be able to relate the number of objects of a given set to a numerical symbol in the form of an Arabic numeral. However, they do not appear to have developed or possess an understanding of how these numerical symbols precisely correlate, as they would apparently need a fully developed ToM for this skill. Yet this is not really certain either, because it may be that they have simply not had the motivation to use these numerical symbols, since such symbols do not play a role in their natural ecological and social environment. The only thing that is certain is that a ToM is apparently a development-related prerequisite for the understanding of the cardinal word principle or the cardinal number principle, and thus the ability to count.

Whether chimpanzees possess a ToM or not cannot be deduced from their apparent lack of understanding with regard to this counting principle, and I have argued elsewhere for a ToM and especially for at least concrete representations of numbers among chimpanzees (*Pan troglodytes*) and bonobos (*Pan paniscus*) (Krebs 2008, p. 128). Nevertheless a cognitive boundary line seems to be drawn by the cardinal word principle, or the cardinal number principle in the case of chimpanzees, which can only be crossed by humans and which leads to mathematical thinking in terms of relations on the basis of representations of numbers, e.g., in the form of Arabic numerals. Accordingly, the discrete ability to count would be a numerical competence that only humans possess, to say nothing of mathematical thinking in terms of relations, e.g., on the basis of Arabic numerals.

8.2.3 Mathematical Thinking in Terms of Relations

What is mathematical thinking in terms of relations, especially on the basis of Arabic numerals or natural numbers? What are the mathematical relationships between the natural numbers in the form of Arabic numerals? In mathematics, the natural numbers 1, 2, 3, 4, ..., denoted collectively by $\mathbb{N}$, are based on Peano's axioms (e.g., Heuser 1994, p. 34; Kennedy 1973, p. 113). They describe the fundamental relationships which characterize the natural numbers and which the natural numbers maintain among each other:

Axiom 1. 1 is a natural number.

Axiom 2. If a is a natural number, then its successor $a + 1$ is also a natural number.

Axiom 3. 1 is not a successor.

Axiom 4. If two natural numbers a and b are different, then their successors $a + 1$ and $b + 1$ will also be different.

Axiom 5. If a set $\mathcal{M}$ of natural numbers contains the number 1 and if $m + 1 \in \mathcal{M}$ always follows from $m \in \mathcal{M}$, then $\mathcal{M} = \mathbb{N}$.

If one defines $2 = 1 + 1$, $3 = 2 + 1$, $4 = 3 + 1$, and so on, then on the basis of this axiomatic principle, it can be shown that 2, 3, 4, and so on are also natural numbers.

The question that arises is how a human brain can even grasp such fundamental mathematical patterns of relationships.

The first cognitive (representational) step, as seen in the foregoing, comprises representations of quantities which the various forms of perception provide (number sense). In the second cognitive (representational) step, they are then symbolically and approximately processed as quantity symbols in the form of Arabic numerals, for example (number processing). Regarding the third cognitive (representational) stage, there is the possibility of mentally representing the fundamental relationships which characterize and maintain the quantity symbols among each other. Let us take the previously introduced cardinal word principle or cardinal number principle, for example. Relatively independent quantity symbols, which can only be processed mentally in an approximate way, become real discrete numeral symbols that can be processed mentally in a precise way. In other words, there is a three-stage representational processing process (at least), in which the discrete numeral symbols are third-order representations and the further mathematical representations of their relationships must be of an even higher order (see Krebs 2008, pp. 159–165).

Therefore, the discrete ability to count is apparently a cognitive capability which presupposes a ToM enabling the intuitive grasp of the relational connections between the individual quantity representations in the form of Arabic numerals and thus provides the representational foundation for more advanced symbolic formal processing. In other words, mathematical thinking in terms of relations, on the one hand, intuitively grasps the relational connections by means of representations of the relationships, and on the other, processes these connections symbolically and formally, by clearly representing the previously comprehended representations of relationships symbolically and formally.

A more precise investigation of this representational functioning of the mathematical thinking process and its breakdown into a more intuitive and a more symbolic–formal domain has been undertaken in detail elsewhere (Krebs 2008, p. 22), so that here I can focus on the representational transitions of the individual numerical competencies (number sense, number processing, and mathematical thinking in terms of relations) and their possible evolutionary origins. After all, it appears that these numerical competencies do not result from a linear line of development starting with number sense, moving on to number processing, and then on to the ability to count down to mathematical thinking in terms of relations. On the contrary, it looks more like the linear line of development is to be found in the representational processing structure of the (human) brain.

Starting from representations of perceptions — the first-order representations — (especially representations of quantity), the first symbolic representations — the second-order representations — (especially the representations of quantity representations) are generated by the brain. Then the brain forms representations regarding the representational correlations between perceptual representations (first-order representations) and symbolic representations (second-order representations). These third-order representations (especially the individual representations of the respective representational correlation between a representation of a quantity and a corresponding Arabic numeral) ultimately facilitate the ability to count, because a discrete understanding of number occurs through these representations of the respective representational correlation between a representation of a quantity and the corresponding symbolic representation, which can precisely distinguish between and represent the discrete number symbols, and thus includes in particular the cardinal word principle or the cardinal number principle, respectively. Thus the cognitive development, which refers, however, to the representational processing structure of the (human) brain and develops through the respective orders of representation, appears to be linear.

With regard to human numerical competencies, the cognitive development then also appears to run along this line of development, for the simple reason that these representations of quantity are first-order representations, while symbolic representations of representations of quantity are second-order representations (here in the form of Arabic numerals), and the representational correlation between these two (the precisely distinguishable, discrete representation of numbers) is a third-order representation. In the following steps of representation, this then ultimately enables the representation of the representational relationships of the individual, precisely distinguishable discrete representations of numbers, i.e., their patterns of relationships among each other (see axioms 1–5 above). It looks like this representational processing structure, or more precisely, its neuronal correlate, represents the basis for the largely modular organization of the (human) brain, as the modularly organized cognitive abilities, such as number sense and ToM, can be classified in representations (see Sect. 8.3.3).

Following this characterization of the mathematical abilities of modern humans, the question immediately arises of course as to their possible evolutionary origins.

8.3 Evolutionary Origins of the Mathematical Abilities of Modern Humans

Since the mathematical abilities of modern human can be divided into three components, namely number sense, number processing, and mathematical thinking in terms of relations, it also stands to reason that the evolutionary origins and a possible evolutionary function are to be sought respectively in the corresponding cognitive domains of grasping and processing representations of quantity, symbolic representation of representations of quantity, and representation of the representational relationship itself, e.g., between the representation of quantity and the corresponding symbolic representation.

8.3.1 Evolutionary Origins of Number Sense

With regard to number sense, this means that, from an evolutionary point of view, one first has to ask why such an accumulator mechanism has evolved in so many living organisms (including mammals). What evolutionary benefit does such an accumulator mechanism have? If one observes the social and ecological world of modern humans by way of example, one ascertains that individuals constantly have to deal with other individuals (family members, friends, competitors, etc.) and their ecological and economic living conditions (work, home, leisure, etc.). Individuals compare and assess their social and ecological environment so that they can better position themselves, for example, by asserting their own personal interests over those of others. However, to be able to better position themselves, they have to be able to compare, and to be able to compare, they also have to be able to estimate! In social contexts, for example, they have to be able to estimate the quantity of possible competitors and the quantity of possible cooperation partners, and in ecological contexts, for example, to comprehend the quantity of possible sources of danger in a habitat.

An accumulator mechanism represents a cognitive solution to this evolutionary problem of the comparison of quantities. Dehaene also comes to this conclusion in his deliberations (Dehaene 1997, p. 27):

> It is even likely that a mental comparison algorithm was discovered early on, and perhaps even reinvented several times in the course of evolution. Even the most elementary organisms, after all, are confronted with a never-ending search for the best environment with the most food, the fewest predators, the most partners of the opposite sex, and so on. One must optimize in order to survive, and compare in order to optimize.

Therefore, it is not really surprising that this fundamental accumulator mechanism, as described above, can be detected in the brains of rats, parrots, monkeys, the great apes, and humans.

In summary, therefore, it can be noted that number sense — an accumulator mechanism — is a cognitive adaptation (a module) in human and animal brains, which has the evolutionary function of being able to approximately grasp and process the number of elements of a given set in the form of representations of quantity.

What about the processing of numbers, the symbolic representation of representations of quantity? Does this also have an evolutionary function?

8.3.2 Evolutionary Origins of Number Processing

As has already become evident in the foregoing discussion, this cognitive ability of being able to form a first symbolic representation through a representation of quantity, for example, is a fundamental understanding of symbols. So the question is what could be the evolutionary origin of such an understanding of symbols? On the basis of the work of Boysen und Berntson (1989) already presented, it has become clear that chimpanzees are able to deal approximately with symbolic representations in the form of Arabic numerals. What was still not clear in this study, however, was the answer to the question of what benefit chimpanzees could gain by being able to cope with symbolic representations like Arabic numerals?

This important correlation is obvious on the basis of two further studies (Boysen & Berntson 1995; Boysen et al. 1996). In these studies, the authors were able to demonstrate how chimpanzees use their understanding of symbols to optimize problems when faced with situations of social competition. For the selecting chimpanzee, the optimization problem consisted of the task of selecting between two dishes containing different series of chocolate peanuts, which were then given to the observing chimpanzee. The selecting chimpanzee was given the chocolate peanuts contained in the remaining dish. Therefore, the expectation was that the selecting chimpanzee would choose the smaller quantity of chocolate peanuts first, as it would then have to turn them over to the observing chimpanzee, in order ultimately to receive the larger quantity for itself. However, none of the chimpanzees displayed this behavior. On the contrary! On average, the selecting chimpanzee always chose the dish with the larger quantity of chocolate peanuts, even though it then had to give them up.

Only when the chocolate peanuts were replaced by Arabic numerals, with which the chimpanzees were familiar, did the chimpanzees display the expected optimal behavior. Accordingly, the chimpanzees' understanding of symbols enables a more optimal behavioral strategy with regard to optimization problems when dealing with situations of social competition for certain preferred resources, by being able to suppress a certain trigger stimulus of this preferred resource by a symbolic representation. Note in particular that little stones were insufficient as symbols representing chocolate peanuts for chimpanzees to show the behavioral strategy that was optimal for them (Boysen et al. 1996). This implies that the suppression of the stimulus only works with the more abstract numerical symbols.

Since this correlation was also demonstrated in human children and adults (e.g., Mischel et al. 1989; Forzano & Logue 1994), the conclusion can be drawn that this evolutionary function of the understanding of symbols lies in being able to suppress the direct stimulus of the relevant representational reference. In the studies with chimpanzees cited above, this was the corresponding number of chocolate peanuts, i.e., their representations of quantity, which lost their direct stimulus by a symbolic representation in the form of Arabic numerals. This is why I have already argued elsewhere for a distinction between number sense and number processing (Krebs 2008, p. 128). However, there I did not yet discuss the representational transition from the representation of quantity via the symbolic representation of the representation of quantity down to the discrete representation of numbers, which represents the representational relation between the respective representation of quantity and the corresponding symbolic representation with this much precision, as I did not differentiate between Arabic numerals used as a symbolic representation of a representation of quantity and Arabic numerals as a representation of the representational relation between representation of quantity and its symbolic representation.

In this context, the confusing thing is the fact that symbols of numbers in the form of Arabic numerals are used for both the symbolic representation (the representation of a representation of quantity) and for the representation of the representational relationship between the symbolic representation and the corresponding representation of quantity. The difference between the two only becomes obvious, as described in the foregoing, by making a distinction between Arabic numerals used as an approximate symbol of quantity (representation of the representation of quantity) and Arabic numerals used as discrete number symbols (representation of the representational relationship between the symbol of quantity and the representation of quantity). Therefore the studies with chimpanzees discussed here (Boysen & Berntson 1989, 1995; Boysen et al. 1996) apparently always refer to Arabic numerals as approximate symbols of quantity and not to Arabic numerals as discrete number symbols, which presupposes an understanding of the cardinal word principle, or the cardinal number principle, respectively, as discussed above, even though Boysen und Berntson (1989) use the term 'counting', without claiming that chimpanzees can count:

> Her [Sheba] success with the symbolic counting task provides evidence that she can use numbers representationally. She was able to substitute abstract Arabic numerals for things to be counted and functionally sum them. Without positing that Sheba can count, a parsimonious explanation on the basis of simpler numerical operations is improbable, given her test performance.

Note also that this study (Boysen & Berntson 1989) does not prove that chimpanzees do not have an understanding of Arabic numerals as discrete number symbols.

The problem with all studies like this or of a similar nature is the fact that they are highly training-intensive and that a performance close to 100% almost never occurs or is achieved only after a great deal of time has been spent (e.g., Biro & Matsuzawa 2001, Fig. 2). Furthermore, if it turned out that the chimpanzees had actually understood the cardinal word principle or the cardinal number principle, i.e., that the last numerical symbol in a counting chain discretely represents the number of elements of the given set, then they would also have to be able to discretely differentiate its numerical number symbols. But this would mean that, especially after intensive training and after sufficient time for conducting the experiment, almost no distance effect should occur during the experiments, in which a higher frequency of errors for number systems that are close to each other (such as 1–2, 3–5, etc.) is measurable in contrast to number symbols that are more distant from one another (such as 1–8, 3–9, etc.). However, this distance effect can be detected in the case of chimpanzees, especially for a number-arranging task with the numbers (0–9) (Biro & Matsuzawa 2001, Fig. 4).

This tends to indicate that chimpanzees do not have an understanding of discrete number symbols in the form of Arabic numerals which represent the representational relationship between a representation of quantity and its symbolic representation, so that a precise comparison of two symbolic representations becomes possible. At the same time, this means that symbolic representations of numbers, which do not yet need to be discrete, are sufficient for the suppression of a stimulus. Therefore, one cannot also conclude that the evolutionary origin or the evolutionary origins of the ability to count can be traced solely back to this number processing and to number sense. So in the event of the ability to count (especially of the mathematical ability to think in terms of relations), other evolutionary correlations will have to be considered.

In summary, it can also be noted with regard to number processing that this approximate processing of numbers in symbolic form, such as Arabic numerals, is a byproduct of an evolutionary adaptation, namely the understanding of symbols, which in particular can represent representations of quantity symbolically and which has the evolutionary function of being able to suppress a specific stimulus in the context of social competition, such as the competition for a certain number of items, in order ultimately to derive greater benefit.

The question that still has not been answered concerns the evolutionary origin of mathematical thinking in terms of relations, and in particular of the ability to count.

8.3.3 Evolutionary Origins of Mathematical Thinking in Terms of Relations

An important indication for a possible evolutionary origin of mathematical thinking in terms of relations was already provided by the results obtained from Wynn's study on the cardinal word principle (Wynn 1990). As described in more detail above, this study showed that children only have such a numerical understanding, which is crucial for the discrete ability to count, from the age of about 3.5 years. The key feature of these results is the age of the children, since children also have a fully developed ToM from an age of about 3.5 years.

This ToM can also be classified in representations, namely as the ability to use third-order representations. Such third-order representations are designated as meta-representations (Perner 1991, pp. 82–89; Krebs 2008, pp. 92–104). Thus, third-order representations are needed for a ToM and the ability to count. Therefore, the question of the evolutionary origins of a ToM could also help to answer the question of the evolutionary origins of mathematical thinking in terms of relations. First of all, one has to inquire about the evolutionary domain of the ToM module.

Sperber (1994) divides this domain of the ToM module into a proper and an actual domain. In doing so, the proper domain would consist of the attribution of behaviorally effective mental states to oneself and others, and the actual domain would be the sum of all representations on the basis of which an individual concludes or could understand their existence and content. Accordingly, this would comprise all of the preceding first-order and second-order representations. I have already pointed out elsewhere that this division of the evolutionary domain of the ToM module is problematic, because Sperber, with his characterization of the actual domain, basically equates it with the completely representational processing structure in the (human) brain. However, this only represents the basis for the modular organization of the (human) brain, and I have hypothesized that the actual domain of the ToM module comprises not only the attribution of behaviorally effective mental states to oneself and to other subjects, but also the attribution of behaviorally effective mental states to any (fictitious) objects and subjects (Krebs 2008, pp. 102–104). After all, the crucial feature for the activation of the ToM module is the perception of behavior which could come from an intentional source and which can be ascribed to mental states. However, whether or not these are always subjects (other persons or animals) does not play a crucial role here, since (fictitious) objects and subjects (such as computers, cars, mythical creatures, gods, etc.) can also be viewed as an apparent source of behavior.

Does this division of the evolutionary domain of the ToM also provide an indication for its evolutionary origins? The answer is affirmative, because the proper domain of the ToM module points to the evolutionary role of the ToM module during the course of the social intelligence revolution of modern humans. For all individuals that live in social groups, it is important that every individual should be able to find his or her way in his or her social group. This applies especially to humans, and requires that every individual of a social group should be able to recognize the other members of this social group as subjects with their own needs, desires, intentions, and so on, which will have an impact on their actions. And this is precisely what a ToM achieves, i.e., it comprehends mental or intentional states.

Accordingly, the relative point in time during hominization as of when ToM can be demonstrated would be of particular interest. Dunbar (2003) exposed this as part of his 'social brain hypothesis'. In this evolutionary context, Dunbar argues that the selection pressure of the ecological environment during the course of hominization, especially in the form of predation pressure, would have resulted in a steady enlargement of social groups. This in turn, would have led to social selection pressure, i.e., being able to cope as an individual in an ever larger social group, and would ultimately have triggered a steady increase in the size of the neocortex. Dunbar then classifies this steady increase in neocortex size on the basis of achievable intentionality classifications during the course of hominization (Dunbar 2003, Fig. 4). As these intentionality classifications have to be converted by means of the formula of 'nth intentional order $= n+1$ th representational order' (Krebs 2008, pp. 95–96), this leads to the following phylogenetic development process in the light of Dunbar's results:

1. In the case of Australopethecines (*Australopithecus*), a representational ability of approximately the third order.
2. In the case of *Homo habilis*, a representational ability that lies approximately between the third and the fourth orders.
3. In the case of *Homo erectus*, a representational ability of approximately the fourth order.
4. In the case of archaic *Homo sapiens* (*Homo heidelbergensis*), a representational ability that lies approximately between the fourth and fifth orders.
5. In the case of modern humans (*Homo sapiens*), a representational ability of approximately the fifth order.

Do fifth-order representations represent the upper limit for the human brain then? This applies at least to representations that are deliberately processed, i.e., in the working memory (Kinderman et al. 1998, Fig. 1; Krebs 2008, pp. 104–110).

Overall therefore, Dunbar provides not only a phylogenetic starting point for a ToM in the case of humans throughout the whole revolution of social intelligence, but also evidence for the phylogenetic development of the representational processing structure in the (human) brain, which is linear until the *Homo erectus* period and exponential thereafter (Dunbar 2003, Fig. 4). The phylogenetic engine behind the representational processing structure in the (human) brain is, therefore, the social intelligence revolution during the course of hominization. Thus, ToM is an evolutionary adaptation to the social environment and it belongs to the evolutionary environment of the social intelligence revolution.

But how is this related to mathematical thinking in terms of relations? What is the link between ToM, being able to count, and mathematical thinking in terms of relations? In order to be able to clarify these evolutionary correlations, the internal structure of human groups has to be visualized once more. After all, they consist of a complex pattern of relations (who has to do, had to do, or might have to do, with what, with whom, when, where, and how). ToM thus represents the cognitive basis for being able even to intuitively comprehend such a complex pattern of social relations, before it can be processed further as part of a social thinking process.

How this social thinking process can be envisaged more precisely in terms of how it could arise from the evolutionary environment of social intelligence and how it could be classified representationally are questions that I have already examined elsewhere (Krebs 2008, pp. 142–153). The crucial feature of this social thinking process (see Krebs 2008, Figs. 3.2 & 3.3) is the fact that a complex social pattern of relations in which subjects represent the junctions is thereby grasped and processed through representations in the brain. As already explained in the foregoing, however, it does not matter whether there really is a subject behind a perceived behavior or not, with regard to the activation of the ToM module. It could just as easily be (fictitious) objects, (fictitious) subjects, or natural numbers, which apparently show behavior.

Therefore, the ToM module not only offers access to social patterns of relations with subjects acting as junctions, but also to fictitious social patterns of relations (such as they occur in fairytales), to ecological (in particular, technical) patterns of relations, and to mathematical patterns of relations. These patterns of relations differ merely with regard to the nature of their junctions. In the case of social patterns of relations, these are the (fictitious) subjects. In the case of ecological (in particular, technical) patterns of relations, they are (fictitious) objects. And in the case of mathematical patterns of relations, they are mathematical objects, such as natural numbers. Accordingly, the mathematical ability to think in terms of relations is a byproduct of social thinking, which itself represents an evolutionary adaptation to a social environment that is becoming more and more complex (see also Krebs 2008, pp. 153–158). The fundamental mathematical pattern of relations with regard to natural numbers is illustrated by the axioms 1–5 (see Sect. 8.2.3) and the natural numbers themselves are its junctions. The mathematical ability to think in terms of relations is, therefore, nothing else but social thinking in terms of relations in its actual domain.

Note, therefore, that the mathematical ability to think in terms of relations is a byproduct of an evolutionary adaptation, namely social thinking, which intuitively grasps social patterns of relations as representations and can process them as representations. With regard to the mathematical ability to think in terms of relations, it is the mathematical patterns of relations that are intuitively grasped as representations and then processed further as representations. In particular, this mathematical ability to think in terms of relations then also represents the representational correlations between representations of quantities and the respective corresponding symbolic representations, so that precise comprehension and processing of discrete numbers in the form of Arabic numerals is ensured, and then the ability to count also becomes possible.

In particular, the phylogenetic development of the mathematical abilities of modern humans does not appear to be a direct linear line of development, but one that is tied to the different stages of the phylogenesis of a representational processing structure. Although this development is linear until the *Homo erectus* period, it then increases exponentially. Moreover, number sense is only an evolutionary adaptation, whereas number processing and the mathematical ability to think in terms of relations are byproducts of the understanding of symbols or of social thinking, respectively. This illustrates a possible answer to the question of the evolutionary roots of the mathematical abilities of modern humans.

8.4 Discussion: The Mathematical Nature of Modern Humans

If the mathematical nature of modern humans is considered in its entirety, then it appears that one cannot assume, either ontogenetically or phylogenetically, a direct linear development beginning with number sense, moving to number processing, and ending with the mathematical ability to think in terms of relations (especially the ability to count). Instead it appears that the line of development of mathematical abilities in modern humans is linked ontogenetically and phylogenetically to the line of development of the representational processing structure in the (human) brain. And although this is a steady process when measured on the basis of the respective representational orders, it is definitely not linear right along the line, especially with regard to phylogenesis, but is in fact exponential as of a certain point in time (the *Homo erectus* period). Nevertheless, the correlations can be stipulated as far as ontogenesis is concerned (see Table 8.1).

The ontogenetic and phylogenetic starting point here is number sense, the approximate comprehension and processing of quantities, which does not require symbolic representations of quantity or discrete counting. The first symbolic representations of the individual representations of quantity, which can then also be processed further in an approximate way without facilitating a discrete ability to count, however, form the second ontogenetic and phylogenetic step. The first representations of the representational relationship between a representation of quantity and the corresponding symbolic representation form the third step, both ontogenetically and phylogenetically, facilitating a discrete comparison of numbers and thus the ability to count, and ultimately leading to higher mathematical thinking in terms of relations.

Table 8.1 Representational correlations between individual mathematical abilities

	Number sense	Number processing	Mathematical thinking in terms of relations (especially counting ability)
First order (from birth on)	Representations of quantity	—	—
Second order (from 1st to 2nd year)	Representations of quantity	First symbolic representation of representation of quantity	—
≥ Third order (from 3rd to 4th year)	Representations of quantity	Symbolic representation of representation of quantity	First representation of the representational relationship between the representation of quantity and the corresponding first symbolic representation

This representational argumentation thus shows a way to close the explanatory gap between the properties of the natural numbers (a part of the mathematical thinking in terms of relations) and the corresponding representations of quantities. This is why there is no need to seek another basis for number concepts, as proposed by Rips et al. (2008), which is not grounded in representations of quantity, but instead observes the cognitive abilities which appear to play a substantial role in number processing and in mathematical thinking in terms of relations. In view of a direct linear line of development from the representations of quantity to discrete concepts of numbers, Rips et al. (2008) get to the heart of the matter:

> These representations [representations of quantity] may be useful to non-human animals, infants, children, and even adults for certain purposes, such as estimating amounts or keeping track of objects, but they are not extendible by ordinary inductive learning to concepts of natural numbers.

In order to move from representations of quantity to discrete representations of numbers in the form of natural numbers, and finally to the concept of natural numbers (see axioms 1–5), the cognitive path has to be taken via the representational processing structure in the (human) brain. Apparently there really is no direct path through inductive learning.

However, there are also positions which agree in part with the route to explanation for representation presented here. Gelman (Gelman & Gallistel 1978; Gelman 2008) continues to argue that the meaning and function of counting is subordinate to the principles of arithmetic, and hence that the discrete ability to count lies downstream of an understanding of the corresponding principles of arithmetic (such as the cardinal number principle) (Gelman 2008):

> Children understand what counting is about because they can make at least implicit use of the principles of arithmetic reasoning. I think the authors [Rips et al. 2008] seriously misestimate the age [3–4 year-olds] at which children have some nontrivial understanding of cardinality and the successor principle.

As already explained above, children aged 3 to 4 years have an understanding of the cardinal word principle and the cardinal number principle (Wynn, 1990), and for such an understanding, they require third-order representations, which represent the representational relationship between a certain representation of a quantity (e.g., three objects) and the corresponding representation of the number (e.g., 3), so that an understanding of the discrete number 3 becomes possible. Building upon this, children will only then be able to represent the relational correlations of the natural numbers $1, 2, 3, \ldots$, in the form of the principles of arithmetic (cardinal number principle, successor principle, etc.), and ultimately axioms 1–5. And only then can one also speak of a discrete ability to count, with which children understand the relational correlations between the individual natural numbers and are able to represent such correlations mentally.

On the other hand, other partial aspects are once again at variance with the representational route to explanation presented here, such as the contention that subitizing lies downstream from counting (Gelman & Gallistel 1978) and the argument that there was a direct and continuous line of development, both ontogenetically and phylogenetically, with regard to numerical cognition, that ran from discrete and continuous representations of quantity and size directly via real numbers to integers (Gallistel & Gelman 2000).

As far as the first assertion is concerned, subitizing — the representation of small quantities —, as presented in the foregoing, is tied to number sense and is not tied to the ability to count (see in particular Dehaene 1997, pp. 66–70).

As far as the second contention is concerned, Gallistel and Gelman (2000) do actually assume that discrete and continuous representations of quantity or size initially lead directly to real numbers:

> Indeed, the necessity of representing discrete and continuous quantity with a single coherent number system drove the prolonged effort to create the system of real numbers.

And only from these real numbers could the integers, and in particular the natural numbers, be deduced (Gallistel & Gelman 2000):

> Arithmetic reasoning is found in non-human animals, where it operates with real numbers (magnitudes). It may be that evolution provided the real numbers and then getting from integers back to the real numbers has been the work of man.

This position, in which it is primarily the direct conclusion or the direct, continuous line of development from discrete and continuous representations of quantity and size to real numbers and further to integers and then natural numbers, that is problematic, because the concept of real numbers comprises the concept of integers and this in turn comprises the concept of natural numbers, as the natural numbers are a subset of the integers and these in turn are a subset of real numbers.

How is the human brain supposed to develop an understanding of real numbers without first understanding the correlations with natural numbers contained therein? As far as the complexity of the concepts of numbers is concerned, natural numbers form the beginning, and lead the way via the concept of the integers to the concept of rational numbers, and then to the concept of real numbers (see, e.g., Heuser 1999, pp. 48–50). And the preceding discussion illustrates precisely the path from representations of quantity or size, whether it be on the basis of a discrete size (e.g., three objects) or a continuous size (e.g., a certain quantity of materials), via the corresponding initial symbolic representation (e.g., in the form of an Arabic numeral), down to the representation of the representational relationship between these two (of the discrete representation of numbers, e.g., in the form of an Arabic numeral). Naturally, it is a cognitive path, which is continuous but not directly linear, but runs ontogenetically and phylogenetically via the representational processing structure of the human brain. But it is nevertheless a path that can close the explanatory gap between number sense, number processing, and mathematical thinking in terms of relations, and thus facilitates a general evolutionary view of this essential building block of human nature, namely the mathematical nature of modern humans, and can also explain the special epistemological features that result from the phylogenetic and ontogenetic development that characterizes modern humans.

References

Barth H, Kanwisher N, Spelke E (2003) The construction of large number representations in adults. Cognition 86:201–221

Barth H, La Mont K, Lipton J, Dehaene S, Kanisher N, Spelke E (2006) Non-symbolic arithmetic in adults and young children. Cognition 98:199–222

Biro D, Matsuzawa T (2001) Use of numerical symbols by the chimpanzee (*Pan troglodytes*): Cardinals, ordinals, and the introduction of zero. Animal Cognition 4:193–199

Boysen ST, Berntson GG (1989) Numerical competence in a chimpanzee (*Pan troglodytes*). Journal of Comparative Psychology 103:23–31

Boysen ST, Berntson GG (1995) Responses to quantity: Perceptual versus cognitive mechanisms in chimpanzees (*Pan troglodytes*). Journal of Experimental Psychology: Animal Behavior Processes 21:82–86

Boysen ST, Berntson GG, Hannan MB, Cacioppo JT (1996) Quantity-based interference and symbolic representation in chimpanzees (*Pan troglodytes*). Journal of Experimental Psychology: Animal Behavior Processes 22:76–86

Brannon EM, Terrace HS (2000) Representation of the numerosities 1–9 by rhesus macaques (*Macaca mulatta*). Journal of Experimental Psychology: Animal Behavior Processes 26:31–49

Butterworth B (1999) *The Mathematical Brain*. Macmillan, Basingstoke

Call J (2000): Estimating and operating on discrete quantities in orang utans (*Pongo pygmaeus*). Journal of Comparative Psychology 114:136–147

Call J, Tomasello M (2008) Does the chimpanzee have a theory of mind? 30 years later. Trends in Cognitive Sciences 12:187–192

Cantlon JF, Brannon EM (2007) How much does number matter to a monkey (*Macaca mulatta*). Journal of Experimental Psychology 33:32–41

Dehaene S, Dupoux E, Mehler J (1990) Is numerical comparison digital? Analogical and symbolic effects in two-digit number comparison. Journal of Experimental Psychology: Human Perception and Performance 16:626–641

Dehaene S (1997) *The Number Sense*. Oxford University Press, Oxford

Dunbar RIM (2003): The social brain: Mind, language, and society in evolutionary perspective. Annual Review of Anthropology 32:163–181

Forzano LB, Loque AW (1994) Self-control in adult humans: Comparison of qualitatively different reinforcers. Learning and Motivation 25:65–82

Gallistel CR, Gelman R (2000) Non-verbal numerical cognition: From reals to integers. Trends in Cognitive Sciences 4:59–65

Gelmann R, Gallistel CR (1978) *The Child's Understanding of Number*. Harvard University Press, Cambridge MA & London

Gelman R, Butterworth B (2005) Number and language: How are they related? Trends in Cognitive Sciences 9:6–10

Gelman R (2008) Counting and arithmetic principles first. Behavioral and Brain Sciences 31:653–654

Hauser MD, Carey S, Hauser LB (2000) Spontaneous number representation in semi-free-ranging rhesus monkeys. Proceedings of the Royal Society B 267:829–833

Hauser MD, Tsao F, Garcia P, Spelke ES (2003) Evolutionary foundations of number: Spontaneous representation of numerical magnitudes by cotton-top tamarins. Proceedings of the Royal Society B 270:1441–1446

Heuser H (1994) *Lehrbuch der Analysis*, Teil 1, 11. Auflage. Teubner, Stuttgart

Kaufmann EL, Lord MW, Reese TW, Volkmann J (1949) The discrimination of visual number. American Journal of Psychology 62:498–525

Kennedy HC (ed) (1973) *Selected Works of Guiseppe Peano*. University of Toronto Press, Toronto and Buffalo

Kinderman P, Dunbar RIM, Bentall RP (1998) Theory-of-mind deficits and causal attribution. British Journal of Psychology 89: 191–204

Krebs N (2008) *Evolutionäre Ursprünge des mathematischen Denkens*. Logos Verlag, Berlin

Leslie A (1987) Pretense and representation: The origins of 'theory of mind'. Psychological Review 94:412–426

Lipton JS, Spelke ES (2003) Origins of number sense: Large-number discrimination in human infants. Psychological Science 14:396–401

Mandler G, Shebo BJ (1982) Subitizing: An analysis of its component processes. Journal of Experimental Psychology: General 111:1–22

Matsuzawa T (1985) Use of numbers by a chimpanzee. Nature 315:57–59

Meck WH, Church RM (1983) A mode control model of counting and timing processes. Journal of Experimental Psychology: Animal Behavior Processes 9:320–334

Mischel W, Shoda Y, Rodriquez ML (1989) Delay of gratification in children. Science 244:933–938

Pepperberg IM (1987) Evidence for conceptual quantitative abilities in the African grey parrot: Labeling of cardinal sets. Ethology 75:37–61

Perner J (1991) *Understanding the Representational Mind*. MIT Press, Cambridge

Premack D, Woodruff G (1978) Does the chimpanzee have a theory of mind? Behavioral and Brain Sciences 4:515–526

Rips LJ, Bloomfield A, Asmuth J (2008) From numerical concepts to concepts of number. Behavioral and Brain Sciences 31:623–687

Sperber D (1994) The modularity of thought and the epidemiology of representations (pp. 39–67). In Hirschfeld LA, Gelman SA (eds) *Mapping the Mind: Domain Specificity in Cognition and Culture*. Cambridge University Press, Cambridge

Suddendorf T, Whiten A (2003) Reinterpreting the mentality of apes (pp. 173–196). In Sterelny K, Fitness J (eds) *From Mating to Mentality – Evaluating Evolutionary Psychology*. Psychology Press, New York and Hove

Washburn DA, Rumbaugh DM (1991) Ordinal judgements of numerical symbols by macaques (*Macaca mulatta*). Psychological Science 2:190–193

Whalen J, Gallistel CR, Gelman R (1999) Nonverbal counting in humans: The psychophysics of number representation. Psychological Science 10:130–137

Whiten A (1996) Imitation, pretense, and mindreading: Secondary representations in comparative primatology and developmental psychology? (pp. 300–324) In Russon A, Bard KA, Parker ST (eds) *Reaching into Thought – The Mind of the Great Apes*. Cambridge University Press, Cambridge

Woodruff G, Premack D (1981) Primitive mathematical concepts in the chimpanzee: Proportionality and numerosity. Nature 293:568–570

Wynn K (1990) Children's understanding of counting. Cognition 36:155–193

Wynn K (1992) Addition and subtraction in human infants. Nature 358:749–750

Wynn K (1995) Origins of numerical knowledge. Mathematical Cognition 1:35–60

Wynn K (1998) Psychological foundations of number: Numerical competence in human infants. Trends in Cognitive Sciences 2:296–303

Wynn K (1998a) An evolved capacity for number (pp. 107–126). In Cummins DD, Allen C (eds) *The Evolution of Mind*. Oxford University Press, New York and Oxford

Xu F, Spelke ES, Goddard S (2005) Number sense in human infants. Development Science 8:88–101

Chapter 9
Our Way to Understand the World: Darwin's Controversial Inheritance

Michael Schmidt-Salomon

Abstract Shortly after he had completed the first draft of his theory of evolution in 1844, Charles Darwin wrote to his friend Joseph Hooker, the botanist, that publishing the theory seemed to him "like confessing a murder" (Glaubrecht 2009, p. 161). Right from the beginning, Darwin was aware of the far-reaching impact his theory would have. And this was probably one of the main reasons for his postponing the publication of his ideas for such a long time. After the completion of the 230 page text in 1844, it was another 15 years (!) before his famous book *On the Origin of Species* was published. Since that time 150 years have passed, but the theory of evolution is as controversial as ever. Darwin's dangerous idea is still putting many traditional world views through some very hard tests. This is the central theme to which I have devoted the following thoughts. I have divided my study into three parts: I shall start by shedding some light on the conflict between Darwin's challenging idea and traditional (Christian) beliefs, a conflict that has lasted till this very day. In the second part, I want to focus on the ideological abuse of the theory of evolution. The third and final part introduces Julian Huxley's concept of 'evolutionary humanism', which links Darwin's scientific inheritance with a distinctly humanist ethic.

9.1 Knowing Instead of Believing: Why the Theory of Evolution and Traditional Forms of Belief Are Irreconcilable

When Darwin set off on his famous expedition aboard the Beagle, he was still a devout Christian. He states in his autobiography that "many ship's officers laughed at me for citing the Bible as an unchallengeable authority on a certain question of morale" (Darwin 2008, p. 94). At that point in time, as he himself wrote, he was not

Michael Schmidt-Salomon
c/o GBS-Bro Elke Held, Im Gemeindeberg 21, 54309 Besslich, Germany, e-mail: mss@schmidt-salomon.de

U.J. Frey et al. (eds.), *Essential Building Blocks of Human Nature*, The Frontiers   143
Collection, DOI 10.1007/978-3-642-13968-0_9, © Springer-Verlag Berlin Heidelberg 2011

at all willing to surrender his faith. Twist and turn it as he might, what Darwin discovered in nature, however, was simply irreconcilable with what his religion taught him to believe (Darwin 2008, p. 96): "Thus, slowly but surely, a loss of faith took hold of me, which in the end was irrefutable and undivided."

Unlike Ernst Haekel or Thomas Huxley, Charles Darwin was a very cautious man and his works display a remarkable reserve as soon as the ideological consequences of his theory are involved. It was only in his autobiography that he allowed himself to speak more plainly. Here we learn, for example, that the bible "was no more believable than the holy writings of the Hindus or of any Barbarian's religion" (Darwin 2008, p. 95). The more we knew about the irrefutable laws of nature, the more implausible miracles became. So Darwin rigorously rejected all traditional beliefs. As he recapitulates in his memoirs with something of a sense of relief, "nothing is more remarkable than the increase in scepticism or rationalism" in the second half of one's life (Darwin 2008, p. 104). His father had advised him, he says, to keep his religious doubts "meticulously secret", because these doubts "could lead to great matrimonial unhappiness". However, in the second half of his life he knew amongst his few acquaintances some "married ladies who are hardly more adherent in their belief than their husbands are" (Darwin 2008, p. 105).

The world at large did not learn of Darwin's criticism of religion to begin with, thanks to the efforts of his wife Emma, a devout Christian all her life, who lovingly censored her husband's autobiography out of respect for the feelings of religious friends and relatives (Darwin 2008, p. 165). In the end, even that was to no avail. The more the theory of evolution gained ground as a scientific model for explaining the world, the more acerbic were the reactions of those who saw Darwin's discoveries as a threat to their religious convictions.

This reaction is entirely understandable. After all, for centuries people had believed that God had shaped each individual species with his own hands and then provided Man, "creation's crowning glory", with an immortal soul. It was Darwin's word against God's word. For many believers, it was simply intolerable that a dyspeptic English scholar should have the gall to upset their intact world of religion! Soon a large-scale conflict was raging between believers and advocates of the theory of evolution, a conflict which, as we know, is still raging today.

On a global scale, particularly in Muslim countries, but also in those African and South American countries where the Christian faith dominates, and in the USA, considerable majorities in the populations rigorously reject the theory of evolution, or have never even heard of it. In Western Europe, however, the situation looks somewhat better, as we know. The vast majority of Germans, for example, accept *evolution as a fact*, although that does not mean that they would also accept the *theory of evolution* in its entirety. In fact, closer inspection reveals that only very few people would plead the total case for evolution in this country.

To understand this, we need to take a more differentiated view of the anti-evolutionist camp than is commonly the case. Here we find not only the traditional creationists, who take *Genesis* literally, but also advocates of the 'intelligent design' theory and (of particular relevance to Europe) those that favour 'theistic evolution'. To obtain a better idea of the similarities and differences in these concepts, I pro-

Table 9.1 Three forms of creationism

	Traditional creationism	Intelligent design	Theistic evolution
Type	Fundamentalist creationism	Pseudo-scientific creationism	Camouflaged creationism
Content	Even denies 'evolution as a fact' (development of species)	Largely accepts 'the fact of evolution', but contests evolutionary logic	Denies evolutionary logic, especially in regard to the 'higher mental abilities'
Strategy	Religion to take the place of science	Religion should be understood as science	Strict separation of science and religion

pose to label these three variants of the belief in creation as fundamentalist, pseudo-scientific, and covert (disguised) creationism (see Table 9.1).

The easiest to identify is *fundamentalist creationism*. Orthodox Jews, Christians, and Muslims understand the creation myths set down in their 'holy scriptures' as being undeniable and unquestionable declarations of facts, not to be challenged even in detail. Thus, fundamentalist creationists deny evolution as a fact, that is, they deny the gradual development and merging of species and, above all, the descent of Man from primeval primates. Within this creationist camp there are differences in the estimation of the age of the Earth: 'old-Earth creationists' accept the fact that the Earth is already billions of years old (and explain this by arguing that 'divine-creation days' are not human days); 'young-Earth creationists' advocate the particularly crude delusion that the Earth was created at a point in time when the Mesopotamians were already brewing the first beer (Harris 2007, p. 13).

Of course, in view of the overwhelming amount of evidence in favour of evolution, fundamentalist creationism does not need to be taken seriously, either scientifically or philosophically. Politically, however, we do unfortunately have to take it seriously. After all, there are millions of people all over the world who hold fast to beliefs of this kind. This is not only a serious obstacle to scientific understanding, but also poses problems from an ethical standpoint: as a rule people who adhere stubbornly to outdated ways of explaining the world also represent ethical and political positions that are no longer in line with the level our civilisation has meanwhile achieved, e.g., they exhibit homophobia, they are anti-abortion, and they advocate capital punishment.

However, literal interpretations of religious creation myths do not go down very well with the American school authorities or the people of Europe either. This has prompted the creationists to develop new constructs to help them harmonise what they once believed with what they now know better. The most important product to emerge from this strange mixture of genuine faith and half-hearted science is the 'intelligent design' theory, the flagship of the pseudo-scientific creationist movement (Kutschera 2004). Advocates of this theory attempt to integrate findings of cosmology, palaeontology, and evolutionary biology into their belief in the Creation. Although they generally accept evolution as a fact, they question whether the development of species took place as a natural process in keeping with the evolutionary logic of variation, mutation, and selection. Instead, they assume that evolutionary processes were initiated by an intelligent planner using his Creation to pursue specific aims.

Pseudo-scientific creationists attempt to develop a seemingly scientific alternative to the theory of evolution that is compatible with religion. As much as they try to sound scientifically convincing, however, the whole thing has very little to do with real science. This is because the explanatory value of the theory is strictly zero: no predictions can be made and development processes that have already taken place cannot be re-enacted on the basis of this theory.

On closer consideration, even the concept of 'intelligent design' is a matchless absurdity (Schmidt-Salomon 2006). Let us assume, just for the fun of it, that an omniscient, all-powerful God really created the universe, *so that* people can live in it and follow his premeditated plan of salvation. Then we are bound to ask ourselves why, in order to achieve this aim, he made so much pointless effort! How, for instance, can we explain away the fact that this supposedly hyper-intelligent designer first created (a) an immense variety of dinosaurs, then (b) allowed a gigantic lump of rock to land on their home planet, so that (c) the dinosaurs became extinct again, in order (d) to make room for a few rat-size prehistoric mammals, from which the alleged pride of creation, *Homo sapiens*, would develop millions of years later?

How 'intelligent', I ask you, can a 'designer' be, with such an absurd approach?! No graphics agency, however chaotic, no automobile manufacturer, no fashion company, no one who is more or less of sound mind would employ a designer with such a disastrous cost–benefit balance sheet! If we look at our world a little more closely, we see that it is so 'unintelligently designed', so full of "mayhem, misfortune, and mishap" (Voland, personal communication), that the question of believing in an intelligent designer becomes superfluous.

So how does the concept of 'theistic evolution', as espoused by the Catholic Church amongst others, differ from the concept of 'intelligent design'? In terms of content, it is rather difficult to keep the two ideas apart, a situation which occasioned Cardinal Schönborn not a little embarrassment with regard to his pro-intelligent design text in the *New York Times* (5 July 2005). Why is it so difficult to draw a demarcation line? Because, of course, in order to uphold the faith in spite of Darwin, the churches must act on the assumption of a god as creator, a god who wants human beings to be part of his remedial plan, and thus acts as the Spirit rector behind natural phenomena! So the difference between the intelligent design theory of the pseudo-scientific creationists is less a matter of the content of what each believes and more a matter of the communication strategies selected in each case.

Whereas intelligent design theorists strive to use their religious ideas to gain a foothold in scientific research, leading church theologians place a strong emphasis on a strict *separation of science and belief*. According to Pope Benedict XVI, or the Protestant bishop Wolfgang Huber (former chairman of the German Protestant Church), scientists should not express themselves *as scientists* on religious questions and believers *as believers* should interfere just as little in the sciences. Instead, they suggest understanding scientific thinking and religious faith as two sides of the same coin, i.e., as separate, but compatible 'truth systems' entrusted with different duties.

Whereas it is the job of science to explain the world, religion is meant to perform the task of providing an orientation in people's lives.

A division of responsibilities like this may sound sensible at first sight and satisfies our need for harmony. It is, however, only possible on the condition that science and religion really are compatible 'truth systems'. But how should this function? Can one really assume from empirically substantiated theory that Man is a simian-like life form and the chance result of the blind forces of selection, and yet still think that Man was deliberately created by a god proceeding in a well-planned manner? The answer is negative, because this would be a contradiction in itself (see Schönborn 2007, p. 86)! The closer one looks, the clearer it becomes that we must decide: either evolution or creation, clarification or obfuscation, Darwin or Bible, scientific knowledge or religious faith. All attempts to connect the one with the other have failed spectacularly! Even the most profound attempt by Teilhard de Chardin is untenable (see Gould 1998; Wuketits 2009).

So what is to be made of announcements such as the church having accepted the theory of evolution long ago? I can only warn against taking such statements seriously! A closer look is needed in order to understand which aspects of the theory of evolution the churches have accepted and which they have not. Let us take the Catholic Church as an example.

After an interval of nearly a century, during which the Roman Curia apparently hoped to be able to solve the problem of the 'theory of evolution' simply by sitting it out, Pius XII was the first pope to express himself decidedly in public on the theory of lineage. In his *Humani Generis* circular of 1950, he explained that the 'theory of evolution' was a legitimate pursuit under certain circumstances, but he left the question open as to whether mankind had really developed 'in body' from the animal kingdom. Concerning the 'soul' and mankind's so-called 'higher spiritual abilities', the papal standpoint was quite clear: according to Pius XVII, as far as the 'soul' is concerned, the devout Catholic must adhere absolutely to the belief "that it is created directly by God" (Neuner & Roos 1992, p. 205).

Fundamentally, very little has changed as regards the position of the Church on this matter, right down to this very day. The only recognizable progress consists in the fact that the Vatican now acknowledges that the question of the physical origin of mankind as being part of the animal kingdom is no longer an 'open question', but, since John Paul II, has been established as a 'fact'. Even so, any explanation of the 'mental', i.e., the psychic, cognitive, and affective characteristics of *Homo sapiens* in terms of evolution is still vehemently contested by the church, although Darwin also did important pioneering work in this field, as documented by his works *The Descent of Man and Selection in Relation to Sex* and *The Expression of the Emotions in Man and Animals*.

In this respect, it may be said that the Catholic Church has at best made friends with 'half a Darwin'. Although it accepts the fact of evolution as a process of development, i.e., also the physical origin of Man from previous primate forms, it does not accept the evolutionary logic behind these developments, especially in relation to the so-called 'higher, spiritual faculties' of Man. So in this respect, the church can set itself apart from 'fundamentalist' as well as 'pseudo-scientific' creationism.

It must nevertheless argue 'creationalistically' in terms of a belief in the Creation, since the Christian faith, as a faith based on the teaching of 'original sin', 'redemption', and 'resurrection', is possible only if God is pursuing a specific aim for his creation. So church theologians have no choice but to fall back on the fiction of a divine designer, a controller working behind the phenomena, and a planner of evolution, if their faith is not to be more than an aggregation of pious-sounding and empty formulas, devoid of all content.

Just like his predecessors, therefore, Pope Benedict XVI leaves no room for doubt that mankind — Darwin or no Darwin — is a divinely intended creature (Horn & Wiedenhofer 2007, p. 15): "Man's singular status of being known and wanted by God, we call His special creation." The pope made this unmistakably clear in his inaugural sermon, when he categorically rejected the position of evolution theory (Horn & Wiedenhofer 2007, p. 7): "We are not the accidental and pointless product of evolution. Each of us is the fruit of a thought of God. Everybody is wanted, everybody is loved, everybody is needed." And even Germany's Protestant Church still sees in God the mastermind and driving force behind evolution. In its recommendations for religious instruction at school, it reaffirms Luther's conviction, namely (Protestant Church Office Germany 2008, p. 10): "Where God does not begin, there nothing can be or become, where he ends, there nothing can exist."

Both the major churches are bound to adhere to the notion that God, in his omnipotence and omniscience, conceived the Creation from the outset precisely so that we humans had inevitably to originate in his own image. The consequences of this are decidedly odd. Had God adjusted the parameters of evolution only slightly (had he, for example, done without the disastrous impact of an asteroid 65 million years ago), then 'Jesus', his alter ego, would presumably have had to reincarnate as a *Tyrannosaurus rex*, rather than adopting a human form! What kind of cult would have arisen from that is beyond contemplation!

On a more serious note, a closer examination reveals that the churches' putative acceptance of the theory of evolution is, in fact, a sham! In reality, the churches accept only 'half a Darwin', so that in this way they may at least be able to salvage the last remnants of their creationist ideology.

But what if, unlike the churches, we were to dispense with this camouflaged creationism and instead took the philosophy of Darwin and his successors seriously? In this case we could no longer consider *Homo sapiens* as the god-willed crown of well-intended, well-designed Creation, but as an unintended, cosmologically insignificant, and transitory peripheral phenomenon in a meaningless universe (Schmidt-Salomon 2006, p. 24). Just in case you are frightened by this statement, it is not as dismal as it may first appear. However, before I start to explain why, we should first take a look at some dark patches in the history of the theory of evolution.

9.2 Errors and Misunderstandings: The Theory of Evolution and the Ghost of Biologism

As we know from bitter experience, great ideas invite a great deal of abuse. Unfortunately, Darwin's theory of evolution is no exception in this respect. The history of the theory of evolution is pitted with fatal errors and misunderstandings, the aftermath of which is visible even today (not least in the sermons of Catholic bishops warning of what they allege to be the "culturally destructive effects" of "rampant evolutionism"). I should like to subsume the undesirable developments in the context of evolutionary theory under the heading 'biologism', making a distinction between theoretical and normative biologism (Schmidt-Salomon 2007). First a brief explanation of the terms:

- The term 'theoretical biologism' refers to all of those ideological models that attempt to explain human behaviour patterns or social circumstances mainly by reference to biological laws, *without* giving due consideration to the specific characteristics of the human species (in particular the importance of cultural factors).
- As distinct from this, the term 'normative biologism' includes all of those ideologies that spontaneously derive what they think 'should be' in moral and/or political terms, from what 'actually is' in biological terms.

Theoretical biologism and normative biologism do not necessarily go hand in hand, but in the past they tended to appear as a 'package deal'. I would like to demonstrate this by means of a short analysis of normative biologism as manifested in Social Darwinism, racism, and eugenics.

Let us start with Social Darwinism. This term, which would probably have met with strong resistance from Darwin himself, describes a form of normative biologism that misunderstands Darwin's theory, especially the phrases 'struggle for existence' and 'survival of the fittest', as an appeal to create corresponding societal conditions (the key phrase here is 'law of the jungle'). In the past, it was considered to be a characteristic feature of the Social Darwinists to see the willingness to wage wars as an 'immanent human characteristic' and to interpret relationships among states and peoples as a 'struggle for living space'. At present, a different Social Darwinist pattern of argumentation seems to be more important, namely the justification of social inequality by referring to the universally applicable principle of self-interest.

What arguments can be held against this? First of all, ethically speaking, Social Darwinist statements are ill-founded, because they are based on a 'naturalistic fallacy' — just like every form of normative biologism. It is not possible, after all, to simply deduce a 'what should be' from an assumed empirical 'what is' (Schmidt-Salomon 2006, p. 93). We cannot just go ahead and 'distil' our ethical values from our knowledge of nature. For example, it would be absurd to deduce the legitimacy of infanticide from the fact that we come across 'infanticide' so often in nature. But this is not all: even from an empirical point of view, Social Darwinist reasoning is on very shaky ground, because it is not founded on solid scientific knowledge, but on a faulty assumption of theoretical biologism.

In particular, self-interest in nature by no means only implies the assertion of one's own interests at the expense of others, but also the willingness to cooperate, to share resources (Voland 2009). In many cases, animals even sacrifice their own lives to enable others to survive. Thus, in nature, self-interest is not at odds with altruism per se. And if we consider the *particularities* of the human species, the Social Darwinist argument for justifying social inequality becomes completely obsolete. If there is *one feature* that characterises humans more than others, it is their distinct ability to take on a different emotional perspective (see Gopnik 2009, who shows that this empathy can already be found in babies and infants). As humans, we can relate to the sufferings of others. *We literally suffer with them.*

Knowing what an important role pity and empathy play in the emotional experience of *Homo sapiens*, it comes as no surprise to find that many social science studies have ascertained a significant connection between social justice and the feeling of personal well-being (Klein 2002, p. 260). With this in mind, one can only agree with Stephen J. Gould, when he states that the human being has the potential to be a particularly "clever and friendly animal" (Gould 1984, p. 220). Certainly, particular conditions need to be fulfilled before society can develop this potential. The consistent denial of this possibility by theoretical and normative biologists, however, demonstrates how undifferentiated (even in a biological sense!) their picture of *Homo sapiens* is.

Let us turn now to the second form of normative biologism, namely racism: the division of the human race into allegedly 'inferior and superior races' is, unfortunately, an ancient phenomenon in our cultural history. The early theory of evolution finally seemed to provide serious scientific evidence for sustaining this prejudice. Although the theory made it clear that even 'civilised humans' originate from 'primitive archetypes', there was still the impression that some parts of mankind were further developed in evolutionary terms, whereas other parts had come to a standstill at an earlier stage of development. Ernst Haeckel's 'racial hierarchy', which we find mainly in his work *Die Lebenswunder* [The Miracles of Life], is a typical example of this way of thinking. Considering the Aryan racial mania that shook the world only a few decades later, not only Haeckel's stepladder with its 'inferior savages' on the bottom rung and the 'superior civilised people' at the top, but also his bluntly expressed conviction that the value of life of these 'inferior savages' is about the same as that of the 'anthropoid apes', appears shocking today. It is but a short step from such a value system to the 'racial hygiene policy' of the National Socialists, who were all too willing to exploit Haeckel's views on the subject (Wuketits 1998, p. 114).

That 'racism' is not compatible with an equal-rights orientated humanist ethic should be self-evident. But it is no longer sustainable in theoretical terms either (especially considering the results of synthetic evolutionary theory based on molecular biology, which Haeckel could not have known about). It is downright nonsense to deduce human characteristics such as intellectual ability from certain physical features, such as skin colour. Numerous scientific studies have led to the not unexpected conclusion that the genetic and phenotypical differences within a so-called 'race' differ more than the differences between these 'races' (Gould 2002). This is one of the reasons why, in modern scientific research (Cavalli-Sforza & Cavalli-Sforza), and fortunately also in political debates (UNESCO 1995), the term 'race has since been dispensed with.

To come now to the third case of normative biologism, the term 'eugenics' was coined in 1883 by Francis Galton, a cousin of Darwin's. What he meant by this was a social-political concept, the aim of which was to increase the share of positively rated hereditary factors in the population and minimise the negatively rated ones through 'good selective breeding'. This was meant to be made possible by encouraging reproduction among the 'genetically healthy' and the prevention of reproduction among the 'genetically unhealthy'. The reason given to justify the need for such 'artificial selective breeding' was the reduction in natural selection mechanisms within the course of the civilisation process. This purportedly involved Man "becoming increasingly like a domestic animal" (Konrad Lorenz) and gave rise to fears that the gene pool would gradually become impaired. This could only be counteracted, so the argument went, by means of government breeding programmes ('racial hygiene').

Just as with other forms of normative biologism, governmentally prescribed eugenics are to be criticised not only from an ethical point of view (especially as it violates the human rights of self-determination), but also from a theoretical standpoint. Such eugenic concepts could, after all, only be considered reasonable on the assumption of 'genetic over-determinism', something that has long since been scientifically refuted, although it has still an astonishing number of advocates in society. This is just one of many examples.

A few years ago, the German philosopher Peter Sloterdijk gave a lecture entitled *Regeln für den Menschenpark* [Rules for the Human Zoo], which created quite a stir after its publication (Sloterdijk 1999). In his lecture Sloterdjik suggested that the so-called 'humanisation problem' might possibly be solved using the methods of genetic engineering, since philosophy and education had failed. The outcry in the feature sections was enormous. Amazingly enough, the extent to which Sloterdijk, and most of his opponents, had misjudged the true reach of genetic engineering was a theme that was given very little attention in this controversial debate! Because, in truth, there are *no morality genes*. Of course, a human being's ethical conduct is not controlled by any arrangement of adenine, thymine, guanine, and cytosine, the four DNA bases.

It was not necessary at all to get to grips with the complex field of genetics to know this. Common sense would have sufficed. From a genetic point of view, the young adults of the 1960s who rebelled against authoritarian power structures, stood up for a freer form of sexuality, and demanded a critical reappraisal of Germany's past, did not differ dramatically from their parents who, to a shockingly large extent, had become the henchmen and executors of Nazi barbarism. The very same genetic pool that led to the prohibition of so-called 'degenerate art' in the 1930s made possible the triumphal march of symphonies by Mahler and the millions of Rolling Stones albums sold in the 1960s. Such a dramatic change in behaviour in one generation cannot be explained solely by genetic determinism. What was crucial here was the radical change in the cultural, social, and economic situation that characterised the convictions of the 1968 generation in contrast to their parents.

Just for the record, anyone who thinks that allowing the ideological consequences of evolutionary theory to exercise a greater influence will lead inevitably to a strengthening of Social Darwinism, racism, and eugenics is very much mistaken, in two respects. Firstly, the theoretical concepts to which normative biologists have referred in the past have in the meantime been adequately refuted. Secondly, one must not overlook the fundamental difference between scientific explanation and ethical justification. Ever since Max Weber at least, we should expect a general acceptance of the realisation that, although science may be able to *describe* what is, it still cannot *prescribe* what is *supposed to be* (Weber 1985, p. 151). The way in which we deal with the findings of evolutionary theory cannot be derived from evolutionary theory itself (see Voland 2009, p. 24). After all, this is not a question of empirical science but a question of philosophical reflection, to which the theory of evolution as a scientific explanatory model can only contribute *indirectly*, namely by demystifying flawed ideological assumptions of being, for example, the notion of a divine designer proceeding according to plan.

As we can see, the theory of evolution is itself not an ideology, but, as a *litmus test for the truth value of statements about the world as it exists*, it has far-reaching ideological consequences. Indeed, most traditional ideologies, especially those solemn 'high religions', have no chance at all of coming out of the 'Darwinian truth test' in one piece. So let us say it loud and clear, even at the risk of risk of offending people who have religious feelings: 150 years after the publication of Darwin's pioneering book *On the Origin of Species*, the best before date for most religious convictions has finally expired! The fact that apologists for religion so frequently point out errors (long since clarified) in the history of evolutionary theory may be interpreted as a tactical manoeuvre to divert attention from this embarrassing circumstance.

9.3 Huxley's Synthesis: How Evolutionary Humanism Comes into Darwin's Inheritance

The English zoologist, philosopher and writer Julian Huxley was not only one of the 20th century's most important proponents of evolutionary theory, but also a politically committed humanist, who, in his capacity as the first Director General of UNESCO in the 1940s, among other things made a major contribution to international understanding after the disastrous experience of the Second World War. Today, Huxley's work is mainly of interest because it brings together two models of thinking that had previously seemed to be irreconcilably at odds with one another, namely humanism and naturalism. By way of explanation, here is a short definition of the terms.

We usually apply the term 'humanism' (if we do not happen to be classical philologists!) to all those systems of ideas that are, firstly, based on the assumption that Man is actively shaping his world, and are, secondly, striving for a consistent

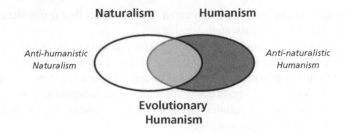

Fig. 9.1 Huxley's synthesis

orientation on individual rights of self-determination, for example, in the sense of the *Universal Declaration of Human Rights*. In this respect, humanism always has a political dimension, because in view of the world's manifold evils, humanists consider an improvement in human living conditions to be not only absolutely imperative, but also possible in principle. Anyone who totally denies our potential to ensure fairer conditions on this speck of dust in the universe is not a humanist, but a cynic.

The term 'naturalism' characterises a basic attitude of the philosophy of science that assumes that everything in the universe has a natural order (Bunge & Mahner 2004, p. 8), in other words, neither gods, nor demons, nor ghosts, nor goblins interfere with the laws of nature. Naturalism says that *everything that is*, is *based* exclusively on *nature*. This means among other things that even Man's so-called 'higher cognitive abilities' are necessarily subject to the causalities of nature, and cannot therefore rise above them. We are the children of evolution — and even the fact that we have invented satellite television or the VW Phaeton cannot alter this.

As already stated, these two different traditions of thinking were brought together by Julian Huxley, who used them to develop a new synthesis, for which he coined the term 'evolutionary humanism' (Huxley 1964). Huxley's synthesis may be represented schematically as in Fig. 9.1. At the intersection of naturalism (left) and humanism (right) we find the new evolutionary or naturalist humanism that was Huxley's concern. On the left, outside the intersection, we see anti-humanist naturalism, which comprises all those naturalist systems of ideas that cannot be brought into line with fundamental humanist convictions. One example is Social Darwinism, already described above, which derives anti-humanist values, for example, in the sense of an alleged 'law of the jungle', from the often cruel survival strategies that have developed within the course of biological evolution.

On the right, we find anti-naturalist humanism, in other words the set of humanist concepts that are at variance with naturalism. Along with religious humanisms that rely on supernatural forces in the fight for better conditions, these also include those secular concepts of humanism that assume the existence of human 'reason' hovering above the physical processes in a 'godlike' manner.

Owing to its naturalist orientation, evolutionary humanism is opposed not only to traditional concepts of God, but also to the traditional idea of Man. It understands *Homo sapiens* (see above) not as the crown of a well-meant, well-made creation, but as an unintentional species of primates created by chance and subject to the laws of nature; one that should not be too conceited about the fact that it has shed its body hair and put on a digital wristwatch.

In my opinion, evolutionary humanism provides an excellent opportunity for coming into Darwin's controversial inheritance. Unlike traditional beliefs, it faces up to the scientific findings which, in the last decades, have led to a thoroughgoing 'demystification of Man'. And yet, in doing so, it sacrifices none of the political visions which, as ethically thinking people, we rightly want to hold on to, for example, the ideas of freedom, justice, and solidarity.

In contrast to Darwin's anti-humanist heirs, evolutionary humanism makes it clear that, on closer examination, the 'ape in us' is not such an unfriendly fellow after all. As already mentioned above, given the appropriate living conditions, *Homo sapiens* certainly has the potential to be a particularly creative, friendly, and humorous animal. What we have to learn, however, is to identify such living conditions more clearly, and also be more resolute about promoting such conditions in the future. Otherwise, the darker side of human nature, which for long stretches of time has turned the history of humanity into a history of inhumanity, is more likely to prevail.

On closer inspection, there is no sensible reason for the fear spread around by theologians that Man would 'become brutalised' if in the future he regarded himself rather as a 'naked ape', since that is actually what he is after all. Quite the opposite! The naturalist 'hominisation of Man' project, in other words, the acknowledgement of our animal nature, can be assumed to be an important precondition for the ethical 'humanisation of Man' project! There would certainly be far fewer tensions in the world if we stopped seeing ourselves primarily as Jews, Christians, Muslims, Buddhists, atheists, etc., but rather as equal members of an ape-like species that tends to enormously overestimate its abilities.

But what about the question of the meaning of life raised in the first part? Is it possible to lead a meaningful existence in a universe which is in itself meaningless? But of course it is! After all, what concerns us when it comes to the question of a meaningful existence has nothing at all to do with (in all probability non-existent) *meaningfulness as such*, that is, the "the great question of Life, the Universe, and Everything", as Douglas Adams once satirically put it (Adams 1981), but merely the far more modest *meaningfulness for ourselves*, in other words, the circumstances that we culturally-shaped primates on a small planet on the edge of the Milky Way may consider to be important to our lives (Schmidt-Salomon 2009).

Being a transient life form on an insignificant planet is at any rate not hurtful per se. It would only be so for someone who asks too much of reality! One may of course be troubled by the realisation that the blind workings of chance and necessity in the universe make it ever so unlikely that there is a loving, omnipotent father keeping a watchful eye over everything. But being allowed to lead one's life *as one sees fit*, without having to subject oneself to a set of principles imposed by an 'imaginary alpha male' may equally well be experienced as a form of liberation. The good news here is that it is precisely the acceptance of the deep metaphysical meaninglessness of our existence that creates freedom to create individual meaning: in a universe that is meaningless 'as such', we can enjoy the privilege of deriving the meaning of life from our own lives.

Anyone who has understood Darwin's inheritance knows that we cannot find the meaning of life *outside of life itself*. Anyone looking for meaning should therefore search for it in the senses, because meaning arises from sensuousness. Two and a half thousand years ago, the Greek philosopher Epicurus stated quite appositely (Laskowsky 1988, p. 98): "I do not know how I can conceive the good, if I withdraw the pleasures of taste, withdraw the pleasures of love, withdraw the pleasures of hearing, and withdraw the pleasurable emotions caused by the sight of a beautiful form."

To learn from Darwin means to understand that there is no such thing as metaphysical good and evil in our world, but most surely a physical weal and woe on which our ethical decisions are based (Schmidt-Salomon & Voland 2008). To say it with Albert Schweitzer (Schweitzer 1974, p. 30): "We are life that wants to live, in the midst of life that wants to live." Time and again, religions and political ideologies have obfuscated this profane fact and driven people into dreadful combat in which *one* mad notion fought against *other* mad notions. Yet another reason why enlightenment is necessary! Of this I am convinced: only beyond good and evil, only beyond the illusions can we realise what life should actually be about, namely the quite profane endeavour to increase happiness on this Earth, and to avoid unhappiness! If, in our ethical decisions, we give due consideration to our own weal and woe as well as other people's, it will be quite simple to find our place as beings in search of meaning in a universe that is in itself devoid of any meaning. To do this we do not require any supernatural sense, no specified plan of salvation, no God — no "lullaby baby from heaven", as Heinrich Heine so neatly put it (Heine 1985, p. 35).

Heine published his poetically ironic criticism of religion in 1844, at the time Charles Darwin was secretly completing his first draft of the theory of evolution. I am convinced that Darwin would have very much approved of Heine's *A New Song, a Better Song*. At any rate it gets to the heart of the message of evolutionary humanism:

Enough bread grows here on Earth,
For all mankind's nutrition,
Roses too, myrtles, beauty and joy,
And green peas in addition.

Yes, green peas for everyone,
As soon as they burst their pods!
To the angels and the sparrows,
We leave Heaven and its Gods.

(Poem translated into English by Joseph Massaad)

References

Adams D (1997) *The Hitchhiker's Guide to the Galaxy*. Del Rey/Random House, New York
Bunge M, Mahner M (2004) *Über die Natur der Dinge*. Hirzel, Stuttgart
Cavalli-Sforza L, Cavalli-Sforza F (1996) *Verschieden und doch gleich. Ein Genetiker entzieht dem Rassismus die Grundlage*. Knaur, München
Darwin C (1993) *The Autobiography of Charles Darwin: 1809–1882*. W W Norton, New York
Darwin C (2008) *Mein Leben. Die vollständige Autobiographie*. Insel, Frankfurt/M
Glaubrecht M (2009) *"Es ist, als ob man einen Mord gesteht" — ein Tag im Leben des Charles Darwin*. Herder, Freiburg
Gopnik A (2009) *The Philosophical Baby: What Children's Minds Tell Us about Truth, Love, and the Meaning of Life*. Farrar, Straus, and Giroux, New York
Gould S (1977) *Ever Since Darwin*. W W Norton, New York
Gould S (1984) *Darwin nach Darwin. Naturgeschichtliche Reflexionen*. Ullstein, Frankfurt/M
Gould S (1981) *The Mismeasure of Man*. W W Norton, New York
Gould S (1996) *Full House: The Spread of Excellence From Plato to Darwin*. Three Rivers Press, New York
Harris S (2005) *The End of Faith*. W W Norton, New York
Harris S (2007) *Das Ende des Glaubens. Religion, Terror und das Licht der Vernunft*. Edition Spuren, Winterthur
Heine H (1985) Deutschland — ein Wintermärchen. In: Heine H: *Werke in zwei Bänden*. Das Bergland-Buch, Salzburg
Horn S, Wiedenhofer S (eds.) (2007) *Schöpfung und Evolution*. Sankt Ulrich Verlag, Augsburg
Huxley J (1964) *Essays of a Humanist*. Harper and Row, New York
Kirchenamt der Evangelischen Kirche in Deutschland (ed) (2008) *Weltentstehung, Evolutionstheorie und Schöpfungsglaube in der Schule*. EKD-Texte, Hannover
Klein S (2002) *Die Glücksformel — oder: Wie die guten Gefühle entstehen*. Rowohlt, Reinbek
Klein S (2006) *The Science of Happiness*. Marlowe, New York
Kutschera U: (2004) *Streitpunkt Evolution. Darwinismus und Intelligentes Design*. Lit, Münster
Laskowsky P (ed) (1988) *Epikur. Philosophie der Freude*. Insel, Zürich
Neuner J, Roos H (eds) (1992) *Der Glaube der Kirche in den Urkunden der Lehrverkündigung*. Pustet, Regensburg
Schmidt-Salomon M (2006) *Manifest des evolutionären Humanismus. Plädoyer für eine zeitgemäße Leitkultur*. Alibri, Aschaffenburg
Schmidt-Salomon M (2007) *Auf dem Weg zu einer Einheit des Wissens*. Alibri, Aschaffenburg
Schmidt-Salomon M (2009) *Jenseits von Gut und Böse. Warum wir ohne Moral die besseren Menschen sind*. Pendo, München
Schmidt-Salomon M, Voland E (2008) Die Entzauberung des Bösen. In: Wetz FJ (ed) *Kolleg Praktische Philosophie*. Bd. 1. Reclam, Stuttgart
Schönborn C (2007): Fides, Ratio, Scientia. Zur Evolutionismusdebatte. In Horn S, Wiedenhofer S (eds) *Schöpfung und Evolution*. Sankt Ulrich Verlag, Augsburg
Schweitzer A (1974) *Die Lehre der Ehrfurcht vor dem Leben*. Union Verlag, Berlin
Sloterdijk P (1999) *Regeln für den Menschenpark. Ein Antwortschreiben zum Brief über den Humanismus*. Suhrkamp, Frankfurt/M
UNESCO (1978) *Declaration on Race and Racial Prejudice*
Voland E (2009) *Soziobiologie — Die Evolution von Kooperation und Konkurrenz*. Spektrum Akademischer Verlag, Heidelberg
Weber M (1985) Die 'Objektivität' sozialwissenschaftlicher und sozialpolitischer Erkenntnis. In Weber M: *Gesammelte Aufsätze zur Wissenschaftslehre*. J.C.B. Mohr (Paul Siebeck), Tübingen
Wuketits F (1998) *Eine kurze Kulturgeschichte der Biologie*. Wissenschaftliche Buchgesellschaft, Darmstadt
Wuketits F (2008) *Evolution ohne Fortschritt. Aufstieg oder Niedergang in Natur und Gesellschaft*. Alibri, Aschaffenburg

Index

U.J. Frey et al. (eds.), *Essential Building Blocks of Human Nature*, The Frontiers
Collection, DOI 10.1007/978-3-642-13968-0, © Springer-Verlag Berlin Heidelberg 2011

THE FRONTIERS COLLECTION

Series Editors:
A.C. Elitzur L. Mersini-Houghton M.A. Schlosshauer M.P. Silverman
J.A. Tuszynski R. Vaas H.D. Zeh

Printing: Ten Brink, Meppel, The Netherlands
Binding: Stürtz, Würzburg, Germany